Beyond Risk Mitigation: Integrating Value Management for Successful Construction Projects

Class

TABLE OF CONTENTS

CHAPTER 1: ORIENTATION

1.1 INTRODUCTION

This chapter provides an orientation to the study applied in the inquiry of the 'Review of risk management tools for construction projects and the implementation of project management strategies' as a contribution to effective project risk management on construction projects aimed at minimising risks which have an impact on project success. The study was undertaken at a school construction project in the Delareyville area within the North West Province of South Africa where the researcher was permanently based as Project Manager. The researcher introduced the context and rationale of the study, before describing the research aims and methodology. A brief description of the ethical measures taken are also presented and followed by the overview of the chapters constituting this study.

1.1.1 *Value Management*

Value management is described as a method used in the construction design process to identify or establish the value system of the client and it also serves to ensure that the client values are communicated and understood by the project team (Green, 1991). Value Management (VM) evolved from Value Engineering (VE), which places significant emphasis on optimising the function of an asset in terms of lowest cost while the VM methodology consists of a number of steps which examine the functional requirements of the client through a process of analysis and then synthesises a suitable design that offers the best value and meets the functional requirements of the client (Thiry, 1997; PMI, 2015). This study endeavored to provide a clear distinction between VE and VM because these terms are mostly applied by different scholars interchangeably. It is for this purpose that the researcher further emphasised that only VM was adopted for use in this study. It is also asserted that Value Management is different from value Engineering in that VM is aimed at the optimisation of solutions in the establishment of suitable decision frameworks for the rest of the project which is achieved through the use of a structured decision model that originates in Multi-Attribute Utility Theory (Green, 1994). A project is a once-off activity, with one chance to achieve a satisfactory outcome and this necessitates the establishment of unambiguous objectives. Value Management identifies the option which gives the best value for money in accordance with the set criteria (Hiley and Paliokostas, 2001).

1

1.1.2 Risk Management

Risk management is defined as the process of identifying and assessing risk and to apply methods to reduce it to an acceptable extent (Tohidi, 2011). This therefore implies that the main purpose of project risk management is to identify, evaluate and control the risk for project success whose overall risk management process includes steps such as, risk planning, risk identification, risk assessment, risk analysis, risk response, risk monitoring and the recording of the risk management process (Baloi and Price, 2003; Alvesson and Ashcraft, 2009; ISO, 2009).

Researchers argue that there is uncertainty in everyday life, in organisations and projects which present a threat to operations of any business or undertaking and the uncertainty in itself is a significant opportunity that must be taken (Hillson, 2011). Construction projects are perceived to be exposed to risk at the time they come into existence and are said to have more inherent risks due to the involvement of many contracting parties such as owners, contractors and designers, among others due to uncertainty in everyday life (El-Sayegh, 2008). A definition from another school of thought describes risk management as a systematic process and contemporary methodology encompassing five stages which are; risk identification, risk analysis, risk evaluation, risk response and risk monitoring (Baker *et al.*, 1999). This study adopted the definition of risk derived from ISO (2009) which describes risk management as steps such as, risk planning, risk identification, risk assessment, risk analysis, risk response, risk monitoring and the recording of the risk management process (ISO, 2009).

1.1.3 Synthesis of Value Management and Risk Management

Value Management and Risk Management are well established disciplines which are recognised as a part of best practice. This is because of their strong link which is premised on the fact that when risk is managed, it is possible to achieve cost-saving and an enhancement in value (Hiley and Paliokostas, 2001). Value Management appears to be one of the best tools for RM because both are compatible and complimentary which provide a potential for a common framework. Risk Management is enhanced by Value Management through the use of the Value Management team to either audit or produce a project's RM plan which further suggests that during the creative stage, alternative methods of mitigating recognised risks are able to be generated (Hiley and Paliokostas, 2001). It is further asserted that the addition of Risk Management also improves awareness of the potential risks of alternative proposals which is a benefit for using the combined approach (Connaughton and Green, 1996; Thiry, 1997).

In essence the narrative of value management which entails getting the same value at the same quality but less cost is derived from the fact that when risk is not managed, it has the potential of negatively affecting project constraints such as bad quality which is only corrected at the expense of time and cost. This further means that when any of the project constraints is not managed, the same product may be attained at a higher cost with compromised quality and outside the time frame (Atkinson, 1999). This is the point at which Risk Management synthesizes with value management in order to realise the same satisfaction at a lesser cost (Mandelbaum *et al.*, 2009).

Research previously undertaken by Bolton (2006) concluded that there is a difference in the standard of infrastructure between rural and urban schools as the quality of the former is relatively inferior to the latter and this is cited as one of the main causes of community protests because those in rural areas want better infrastructure (Bolton, 2006). In order to address this problem, the government initiated an infrastructure development programme aimed at bridging the infrastructure gap. A rollout of construction projects to build new schools and renovate existing dilapidated schools was therefore implemented (Bolton, 2006).

It is worth noting that the South African government procurement system of capital infrastructure projects is managed through the National Department of Public Works, which is also constitutionally mandated to provide cost effective solutions (Bolton, 2006). This means that the department plays a leading role in risk management related to construction projects, where float and contingencies are the frequently used safety nets to mitigate construction related risks (Kaseke, 2011). Other risk management techniques apart from contingencies and float are, Fuzzy Sets Theories (FST), Functional analysis systems technique (FAST) Models, Decision Trees, Brainstorming, Case Based Approach, Checklists (PMI, 2015; Lyon & Skitmore, 2004).

Further, additional techniques used for project risk management have been identified by scholars and these include techniques such as, the Simple multi-attribute rating technique (SMART), Kano model, Lever of value, Quality function deployment technique, Redress,

Special adjacent programme (SAP), Time, cost and quality triangle, Value analysis (VA) and Life-cycle costing (Bowen, Jay, *et al.*, 2010; Hayles *et al.*, 2010; Bowen *et al.*, 2011).

Research undertaken on the operations of the National Department of Public Works revealed that many projects face similar problems (Kaseke, 2011). The identified problems include, among others, delayed completion time, budget over-runs, abandoned projects, poor workmanship and premature termination of contracts by the client due to non-performance (Kaseke, 2011). These problems are attributed to lack of a reliable tool or technique to use as an effective project management strategy for risk management on construction projects during the planning and subsequent project phases (Kaseke, 2011). In addition, it is also noted that there is a lack of reference to databanks to ensure that lessons learnt from past projects are transferred and utilised on future similar projects (Bowen, Cattell, *et al.*, 2010). This observation is also supported by guidelines provided by the Project Management Body of Knowledge in terms of project closure which outlined lessons learnt as a key activity of that process (PMI, 2015). The Value Management methodology also emphasises the importance placed on documenting project data because organisations effectively improved project performance and increased project management competency by learning from previous project successes and failures (Rowe, 2008).

The present study also explored the application of the Integrated Approach for Value Management as a contributor to effective project risk management for construction projects. Although the term Value Management is sometimes used interchangeably with Value Engineering, this study adopted the use of Value management as the preferred terms for the purpose of this study will assume the definition assigned to it as a systematic approach directed at analysing the function of systems and supplies for the purpose of achieving their essential functions at the lowest cost and time consistent quality (Mandelbaum, 2006; Mandelbaum *et al.*, 2009).

1.2　BACKGROUND TO THE RESEARCH

The slow delivery of school infrastructure is cited as one of the reasons for the huge backlog of school infrastructure faced by many communities especially in the rural areas of South Africa. The above assertion is based on previous studies which suggest that a number of school projects are unable to be successfully completed for use by communities where these projects are located (Baloyi and Bekker, 2011). In an effort to address this problem, several school infrastructure development programmes have been initiated by the government and private sector, which includes the construction of new schools and also the upgrading of existing dilapidated school infrastructure, referred to as renovation projects (Bolton, 2006). It is suggested that construction related problems such as contract scope changes and other unforeseen risks related to time, cost and quality are among the main causes of the slow pace at which school projects are successfully completed which has ultimately led to the backlog (Baloyi and Bekker, 2011).

The traditional view of project success criteria places strong emphasis on the dimensions of cost, time and quality (Pinto and Slevin, 1988). Currently, the common measurement of project success is on the basis of completing such an endeavour within time, cost and quality as every project is faced with some form of project constraints which impact on time, cost and quality (Walker, 2003). Like any other project, construction projects are not immune to general problems affecting projects and overcoming these problems is determined by the successful interaction of multiple disciplines whose ultimate aim is to provide best solutions and answers from a global and diverse team (PMI, 2015).

A project is a short term endeavour which has a start and end time, meaning that it takes the form of a temporary characteristic trait with processes which evolve through different stages in order to get to the final product (PMI, 2015). The success criteria of any project is therefore largely dependent on the way functionaries respond to problems related to project constraints which have an effect on time, cost and quality. While there is vast knowledge, tools, techniques and models within the project management domain available to functionaries for use as solutions to such problems, the choice of potential solutions for specific problems requires careful analysis so that appropriate solutions are applied to specific problems as there is no tool, model or technique which is a generic solution to all problems, i.e, 'the one size fits all' narrative. In view of this, scholars suggest that project functionaries have to be equipped with abilities to be able to identify and categorise problems affecting a project in order to apply the appropriate solutions because knowledge and know-how are aspects which deal with the processes of learning, understanding and application of information (Soo *et al.*, 2002).

The research described in this book focused on the 'review of risk management tools and the implementation of project management strategies' as a contributor to effective 5 project risk management. One of the strategies considered was the application of the integrated approach model for Soft Value Management as a project management strategy in the management of risk (Pemsel and Wiewiora, 2013). This approach was adapted for use and tested in the risk management of constraints pertinent to construction projects in form of time, cost and quality, which also rank high among common problems affecting projects (Zhi, 1995; Assaf and Al-Hejji, 2006).

Soft Value Management is a concept developed through action research to determine a client's needs and requirements during the project and the process involves capturing requirements and identifying the values of stakeholders in order to avoid the escalation of any of the project constraints (Al-Yami, 2008). The concept is also defined as a form of Value Management which specifically addresses 'soft' problem situations and which also uses an enabling and learning' facilitation methodology (Barton, 2000). Value Management through action research is further defined by Barton (ibid) as a structured facilitated process in which decision makers, stakeholders, technical specialists and others work collaboratively to bring about value-based outcomes in systems, process, products and services.

When a sector formulates a new strategy, one of the key considerations is its ability to exploit opportunities which enable that industry to achieve well defined goals and this involves initiating projects which add value to the industry (Desouza and Evaristo, 2004). This implies that projects have become a game changer in the business world with a shift from the traditional hierarchical structure to a more self-managing setup and the challenge for the industry is to set up a standard model which deals with generic risk challenges for project constraints. The traditional methods of risk management on construction projects such as float and contingencies though still in use, are outdated and do not add the desired value to project success (PMI, 2015). Even though the effect of Value Management is evident and formally recognised, the technique is often neglected and rarely used for risk management. Scholars assert that information about project management processes, best practices which include risk management and project results, are written down and stored but only become valuable information when the techniques are available for use by project team members for practical application (Pemsel and Wiewiora, 2013).

The important factor is to have the different risk management models available to project functionaries and integrated into the overall project charter during the initial stage of the project. In order for this to be realised, an analysis of the current construction risk management techniques needs to be reviewed within the project setup in order to get a clear picture of the workable solutions available to the industry for use by project team members. The result of the analysis will generate a problem solving process which helps to address the needs identified and one such technique which meets this requirement is the 'Integrated Approach Model' for Value Management which is one of the project management strategies.

The Integrated Approach for Value Management adapted from the work of Al-Yami (2008) is a graphical tool. It is used for illustrating the application of this technique as a risk management model and finds itself common to both Value Management and Action Research methods (Al-Yami, 2008). Scholars assert that Value Management techniques are a useful tool in the mitigation of construction related problems on projects yet they are rarely implemented (Bowen *et al.*, 2010). This suggests that many people in the built environment have no knowledge of the vast project management techniques for risk management available while the few who possess the necessary skills and experience to effectively use the techniques to deal with problems inherent to construction projects rarely apply this knowledge on their respective projects.

Studies previously conducted concluded that many risk management tools currently in use, such as float, contingencies, Fuzzy Sets Theories (FST), decision trees and many others have limited success as they are only effective on some projects while they prove ineffective on others hence, the lack of versatility for application to all projects (Raz and Michael, 2001). This is because of the different and unique dynamics of each individual project arising from the fact that every project is unique with individualised risk profiles that require customised risk management strategies (Raz and Michael, 2001). In order to fully understand risk, a suggestion to conduct risk management in two phases is proposed. The first phase is a process consisting of four main phrases which are, risk assessment, risk identification, risk analysing and risk prioritising while the second phase consists of another four processes called risk control, risk management and planning as well as risk tracking and corrective action (Raz and Michael, 2001).

Other scholars propose a risk management style where the identified risk management approach is a multiple phased risk analysis which covers identification, evaluation and control of risks (Tummala and Burchett, 1999). It is further asserted that risk management processes are consistent and structured approaches to enumerating and understanding potential risk factors which leads to the assessment of consequences and uncertainties associated with these identified risk factors (Tummala and Burchett, 1999). Contrary to the multi-phased analysis is the view of other researchers who assert that risk management is a formal orderly process for systematically identifying, analysing and responding to risk events throughout the life of a project to obtain the optimum or acceptable degree of risk elimination or control (Al-Bahar and Crandall, 1990). Further, scholars argue that the best way to deal with risk is to compare different theories used for dealing with uncertainty within the construction industry and recommended Fuzzy Sets Theory (FST) as a solution for assessing construction uncertainty (Baloi and Price, 2003). In order to provide the reader with an insight of how Value Management is synthesised with Risk Management, it is important to provide an understanding of what the two components entail as is provided in the succeeding discussion.

Scholars suggest that most role-players in the sector especially contractors are unable to successfully complete projects on time due to various problems classified as industry or project risks (Akintoye and MacLeod, 1997). Research previously undertaken by Norton and McElligot (1995) reveals that the benefits of knowledge transfer which includes the passing down of appropriate risk management tools and techniques for construction projects has long been recognised in project-based organisations. This process is however frustrated because the models available are not documented to be accessed for use on future projects (Norton and McElligott, 1995).

The complexity encountered in identifying and capturing an appropriate project management strategy model for risk management in construction projects makes it very difficult for the industry to develop a standard model for risks associated with project constraints. As a result, this study attempted to apply processes adapted from the Integrated Approach to Soft Value Management as a project management strategy for the purpose of risk management on the construction project as outlined by Al-Yami (2008). The project management strategy for risk management using the Integrated Approach to Value Management during the tenure of a project is very critical in order to monitor and manage potential risks because it incorporates the VM processes at all phases of the project and increases value towards the implementation of Risk Management. The main objective of the proposed framework referred to above is to provide a value methodology for consideration in the use of risk management principles for construction projects by decision makers. This methodology defines terms, establishes its essential elements and principles and also clarifies roles and responsibilities of stakeholders with a view to increasing the value of risk management application in the project (Pemsel and Wiewiora, 2013). It also describes the functions and activities that need to be undertaken at each stage as well as the tools and techniques required to effectively enable clients to satisfy their needs and requirements. In addition, it integrates the data and information obtained from the findings from the action research to realise an approach to help in the implementation of value management to overcome barriers identified in the study (Kelly *et al.*, 1998).

It should also be noted that decades of Bantu Education in South Africa deprived some citizens of viable educational opportunities because it restricted opportunities for the acquisition of technical and professional skills by black people (Jansen and Christie, 1999). This might explain the reason why black contractors are the majority in the category of firms mostly faced with the scourge of failure to achieve project success due to challenges encountered on construction projects (Ngwenya, 2007). Also related to this is the risk of awarding construction contracts to some 'Black African' contractors who are unable to raise the required capital due to their inability to provide financial guarantees on multi million rand projects which has an effect of their cashflow (Ngwenya, 2007; Baloyi and Bekker, 2011). Financial guarantees on multi million rand projects are used by government to recover financial losses incurred when a contractor defaults during the tenure of the project such as delayed completion, poor quality and when the contractor abandons the project.

The fact that property is used as collateral by financing institutions to finance potential borrowers to fund businesses or projects as start-up capital such as construction projects, the drastic curtailment of property ownership rights of blacks during the pre-democracy era in South Africa made it impossible for this category of people to acquire assets. The concept of property rights only applied to all citizens after 1994 resulting in the inability of Black contractors to acquire finance where assets serve as collateral (Jansen and Christie, 1999). This pre-democracy restriction led to some construction projects awarded to 'Black' contractors either being abandoned or terminated prematurely by the clients due to lack of financial capacity to sustain operations. Lack of property ownership inevitably leads to limited liquidity for African contractors.

The fact that little training was available to black people also placed them at a skills disadvantage and made it difficult for them to acquire necessary skills and experience to manage businesses successfully (McGregor *et al.*, 2017). It can also be concluded that the lack of skills and professional abilities of Black contractors to effectively manage construction projects such as lack of financial management skills resulted into bankruptcy and liquidation of such companies and ultimately led to the abandoning of projects or termination by the client (Ngwenya, 2007). Technical skills and finance are therefore two most distinct barriers to success for African-owned construction companies.

In order to bridge the gap between the previously disadvantaged masses and the minorities who were previously advantaged, the government introduced a policy of 'Broad Based Black Economic Empowerment' (BBBEE) aimed at providing preferential procurement of government business to enterprises that are owned or managed by black people (Ngwenya, 2007). The newly enacted legislation of BBBEE served as a turning point in the construction sector as many black owned construction companies came into existence (Ngwenya, 2007). The introduction of BBBEE emanated from the lack of financial capacity among black South Africans and was aimed at providing an empowerment vehicle for previously marginalised individuals (Jansen and Christie, 1999). The provision of empowerment was aimed at making it possible for African contractors to participate in major income-generating activities in the country.

Studies conducted on BBBEE companies revealed that most of the businesses lack resources such as finance, material, and human skills to enable them deal with the problems associated with the construction industry hence the problems of the huge school infrastructure backlog (May and Govender, 1998). These issues became apparent when new companies were awarded construction related tenders for either new construction projects or projects to renovate existing dilapidated schools. Most of these contractors completed the projects beyond the scheduled time, delivered projects of an inferior quality or failed to complete the projects (May and Govender, 1998).

The problems identified by scholars in the above paragraph have been summarised and categorised into potential risks which mainly relate to the project constraints of time, cost and quality upon which this study is focused in term of tools to manage such risks as well as the implementation of an integrated approach or value management as a project management strategy to manage such risks. It is further noted that some projects are either abandoned or have their contracts terminated prematurely by the client as a result of the various construction related risks which include poor quality, delayed completion and budget overruns (May and Govender, 1998). A third reason provided by some scholars is the inability of contractors to duly complete the construction work because of either financial mismanagement or a lack of expertise to correctly price the Bill of Quantities (BoQ) in tender documents (Baloyi and Bekker, 2011). The four major problems noted are just a few of the numerous risks inherent to the construction sector in South Africa currently, which is a possible explanation of the backlog on infrastructure, a subject explored by this study.

Risk management has been identified as a central issue in the planning and management of any venture and the construction industry is found to be vulnerable to more risk and uncertainty than other industries (Enshassi and Mosa, 2015). Further emphasis is placed on the process of taking the project from conception, completion and into use as being a complex issue which needs to be well managed as risk in construction is the object of attention because of time and cost issues related to construction projects (Jaafari, 2001; Enshassi and Mosa, 2015). It is for these reasons that a gap was been identified and the need realised to find a project management strategy to be employed in the management of risks on projects faced with the challenges identified in the preceding discussion.

1.3   SIGNIFICANCE OF THE RESEARCH

The backlog of school infrastructure in South Africa has caused a significant imbalance on existing schools and demand for such infrastructure has increased tremendously. South Africa is currently experiencing a construction boom in order to address the backlog but much of this effort is being frustrated because most contractors are faced with risks which prevent them from completing such projects on time within the cost and at the required quality. The studies previously conducted established that some of the reasons behind the noted problems included the lack of a standard risk management model which would be applied to manage more than one risk at a time. This is hoped to be achieved by exploring the implementation of an integrated approach to Value Management at the strategic level of a project.

The principles and techniques of Value Management provide the required quality at optimum whole life during the process of developing a project, which will provide best value from a whole life perspective. This approach also aims to upgrade Value Management to continue its competitiveness, enhance its performance and spread its implementation through all the phases of the project in delivering value for money. Further, it endeavours to improve and promote the technical knowledge and awareness of Value Management. It is also worth noting that other surveys previously conducted on Value Management with specific reference to the construction industry professionals concluded that although the knowledge about Value Management is common, it was not practised and where it is used, it is merely for the purpose of cost reduction (Bowen, Edwards, *et al.*, 2010). The researchers also note that float and contingencies are mostly included in contract documents to mitigate risk which do not holistically provide a solution to the risks prone to the South African construction sector (Bowen, Edwards, *et al.*, 2010).

In view of the above, this research also endeavoured to analyse the prevalence of value management knowledge among role players in the construction industry within the South African context and also its place as a risk management technique on construction projects. It is also worth noting that information on the risk management tools and techniques currently in use such as float, contingencies, FAST models and FST sourced through literature, case studies and existing published research indicate their inefficiencies as they are partially achieving the purpose of complete risk management on construction projects. It is also suggested that the most frequently used tools for identify risks are, brainstorming, case-based approach and checklists while the most frequently used tool for risk identification is Brainstorming (Lyons and Skitmore, 2004).

The current study also evaluated the application of the Integrated Approach for Value Management as a project management strategy for risk management on the school construction project. This is because it is asserted that the use of traditional risk management tools noted in the previous discussion are not bearing desired results as confirmed by the fact that the construction sector keeps experiencing problems ranging from delays, budget over-runs, and abandoned or termination of contracts despite the availability of numerous risk management tools which also include lessons learnt from previous projects (Baloyi and Bekker, 2011). The researchers' intention is to share the results of this study with the Department of Basic Education, in its capacity as client through the implementing agent, the IDT, which is the employer in order for the recommendations to be adopted as one of the project risk management strategies for the effective management of risk to improve project success rate on construction projects.

1.4   OVERVIEW OF RISK MANAGEMENT TOOLS

The traditional risk management tools and techniques currently in use which are believed not to be providing a holistic solution on construction projects are, Float, Contingencies, Brainstorming, Case Based Approach (CBA), Checklists, FAST Diagram and Fuzzy Set Theories (PMI, 2015). In addition, more others techniques and tools for risk management such as Decision trees, Simple multi-attribute rating technique (SMART), the Kano model, lever of value, quality function deployment technique, Redress, Special adjacent programme (SAP), time, cost and quality triangle, Value Analysis (VA) and Life-cycle costing (Bowen, Jay, *et al.*, 2010; Hayles *et al.*, 2010; Bowen *et al.*, 2011). The intention of the researcher was to outline the tools for the benefit of the reader although this study will not discuss all of them.

1.5   RESEARCH PROBLEM

Risk is an integral part of any project and an appropriate intervention in terms of risk management techniques is very critical. Currently, float and contingencies are provisions incorporated into contract documents to mitigate any kind of potential risk encountered by the project and Fuzzy Set Theories are also used. This makes it impossible to isolate one that may be assumed to be a generic or standard risk management technique capable of addressing majority of risk issues in the construction industry (Horlick-Jones, 2005; Zeng *et al.*, 2007). Exploring the application of the Integrated Approach for Value Management technique presents a potential standard model as a solution to the stated limitation.

When the technique is packaged, analysed and properly documented, it will be used as a potential standard model for risk management within the construction industry in South Africa and serve as a solution to risks inherent to the industry. Once this model is certified, risks which account for the current problems on school construction projects such as poor quality workmanship, budget overruns and lack of capital to effectively sustain projects which ultimately lead to abandoning of projects or termination by the client will be detected and managed in the early phases of the project. The problem investigated in this study is;

School construction projects encounter problems which impede their successful completion within time, cost and quality resulting in a huge backlog of school infrastructure. What can be done to remedy this situation?

1.6   RESEARCH QUESTIONS

In order to fully address the above problem statement, the central research question is:

What can be done to effectively manage construction problems in order to improve the project success rate and reduce the school infrastructure backlog?

As a follow-up to the main research question, the sub questions are;

1.6.1   Sub-question 1

What are the methods of procurement for school construction projects?

1.6.2   Sub-question 2

What are the threat risks faced by each stakeholder?

1.6.3   Sub-question 3

How are these risks currently managed?

1.6.4   Sub-question 4

How effective is the current project risk management?

1.6.5   Sub-question 5

How can Value Management be used in the project risk management?

1.6.6   Sub-question 6

How effective would a Value Management contribution be to Project Risk Management?

1.7   RESEARCH AIM AND OBJECTIVE

1.7.1   The aim of the research is:

To find out if project management strategies can be used on construction projects to improve project success within the success criteria of time, cost and quality.

1.7.2 ***The objective of the research is:***

To improve the project success rate based on the identified project success criteria and deliver projects within time, cost and quality to prevent school infrastructure backlogs.

1.8 SCOPE OF THE RESEARCH

The scope of this study was restricted to the 'Review of risk management tools and the application of Project Management Strategies' on a construction project in the Ngaka Modiri Molema District Municipality within the North West Province in the Republic of South Africa. In view of the fact that project dynamics are not typical of the whole country, the findings may not be generalised to the whole nation.

1.9 LIMITATIONS OF THE STUDY

The single case design is a limitation as it has a tendency to constrain the outcomes which have a tendency to be subjective and based on the dynamics unique to this project. Therefore, the results of this study are only limited to the construction project site being investigated and may not be generalised and applied to other construction sites.

1.10 ORGANISATION OF THE RESEARCH

The research was organised in the following manner:

Chapter 1: This chapter dealt with the introduction and background of the research. Also discussed in this chapter was the problem statement, rationale of the research, the research environment, definition of concepts and how the research was structured.

Chapter 2: Chapter 2 presented the literature review. This formed the literature and theoretical foundation of the research. It was in this chapter where various literature on the subject including published journals and text books were consulted, compared and contrasted. This was done in order to provide a theoretical framework and a perspective of what other writers have previously written on the subject and also analysed the different views expressed by the various authors.

Chapter 3: The chapter consisted of the research design, data collection methods, data analysis techniques and sampling techniques.

Chapter 4: The chapter presented the results and interpretations of the study.

Chapter 5: This was the final chapter which presented the conclusion and recommendations.

1.11 SUMMARY

This chapter discussed among others, the introduction where the main focus of the research conducted was explained, as well as the presentation of background and importance of the research. The problem statement and the research methodology were also presented in this chapter which also included objectives and scope of the research. This was followed by definition of concepts and the organisation of the research. The next chapter focuses on the literature review in which various literature, including books, magazines, journals, research papers, the internet, newspapers and many more were consulted, contrasted and analysed in order to build a theoretical framework of this study.

CHAPTER 2:   LITERATURE REVIEW

2.1   INTRODUCTION

The purpose of this chapter is to reflect the literature reviewed for this study and present the theoretical perspective of the research as well as empirical studies recently conducted which are relevant to the research. Scholars assert that whatever its scale, any research project will necessitate reading what has been written on the subject in a critical review which demonstrates some awareness of the current state of knowledge on the subject, its limitations and how the proposed research aims to add to what is known (Gill and Johnson, 2010). This is the basis upon which this chapter is premised and a further discussion of the theories and empirical studies within the broader research field in general and under sub-headings within the context of Value management and Risk management in particular. A detailed review of literature on Value Management, Risk management, Project success criteria is also presented because studies previous undertaken on this subject suggest that there are no consistent criteria in the interpretation or methodology which is accepted for measurement hence the need to agree on the criteria at project inception (Dube, 2016).

Project success has historically been defined as the ability to meet the expectations of the customer irrespective of whether the customer is internal or external and the criteria for success must be agreed at the commencement of the project to avoid a difference of expected outcomes amongst stakeholders (Kerzner and Kerzner, 2017). The fact that project success implies different things to different people implies that success is dependent on the context in which such people perceive success. A 'one-size-fits-all' instance does not work in this instance because projects are different in terms of risks involved, project dynamics, size, complexity and other variables (Shenhar *et al.*, 2001; Chan and Chan, 2004; Jugdev and Müller, 2005). It is further asserted that some of the aspects upon which project success is measured are safety in use; guaranteed employment, profitability, compliance with regulations and completion within time, cost and quality (Jugdev and Müller, 2005).

2.1.1   Deductive research

It is critical to note that although the concept of deduction as an approach to research is discussed comprehensively in chapter 3, the researcher found it important to address the theoretical basis of this approach in the current chapter. A deductive research method entails detailing the development of a conceptual and structure prior to its testing through empirical observations (Ghaye *et al.*, 2008). This is an approach where a researcher proceeds from general rules of logic (Gill and Johnson, 2010). For the purpose of this study, the general rules of logic on the school construction project under investigation emanated from the view that the project is experiencing risks due to the lack of implementation of effective project management strategies during the planning phase. It is from this conclusion that the researcher explored other project management strategies for risk-inherent activities to be implemented in order to observe and evaluate the effects.

The deductive approach enables the researcher to organise grounds into patterns that provide conclusive evidence for the validity of a conclusion (Palys, 2003). This approach was adopted for the current study as the researcher organised grounds into patterns in which the strategies of project team members were organised into patterns and observed in order to draw conclusions. Additionally, a deductive research method entails the development of a conceptual and theoretical structure prior to its testing through empirical observations (Gill and Johnson, 2010). Similarly, during the current research, the concept of value management being a contributor to effective project risk management, which was conceptualised with structured theory before being tested through empirical observations was employed.

2.2 RISK MANAGEMENT

Risk is defined as a multifaceted concept which has the probability of a damaging event occurring in the project, which has the potential to affect project objectives (Yu, 2002; Wang *et al.*, 2004). Although risks usually have negative results which lead individuals to only consider the negative side of such risks, not all risks are associated with negative results, however, risks also present opportunities (Baloi and Price, 2003; Hillson, 2011). Other scholars define risk management as the process of identifying risks and applying methods to reduce them to an acceptable extent (Tohidi, 2011). Accordingly, the main purpose of risk management is to identify, evaluate and control risks for project success. The overall risk management processes include, risk planning, risk identification, risk assessment, risk analysis, risk response, risk monitoring and recording of the risk management process (Baloi and Price, 2003; ISO, 2009; Lee *et al.*, 2009).

A conceptual link between the principles, framework and process of risk management has been developed and it provides a synthesis between the three components as presented in figure 2.1 below.

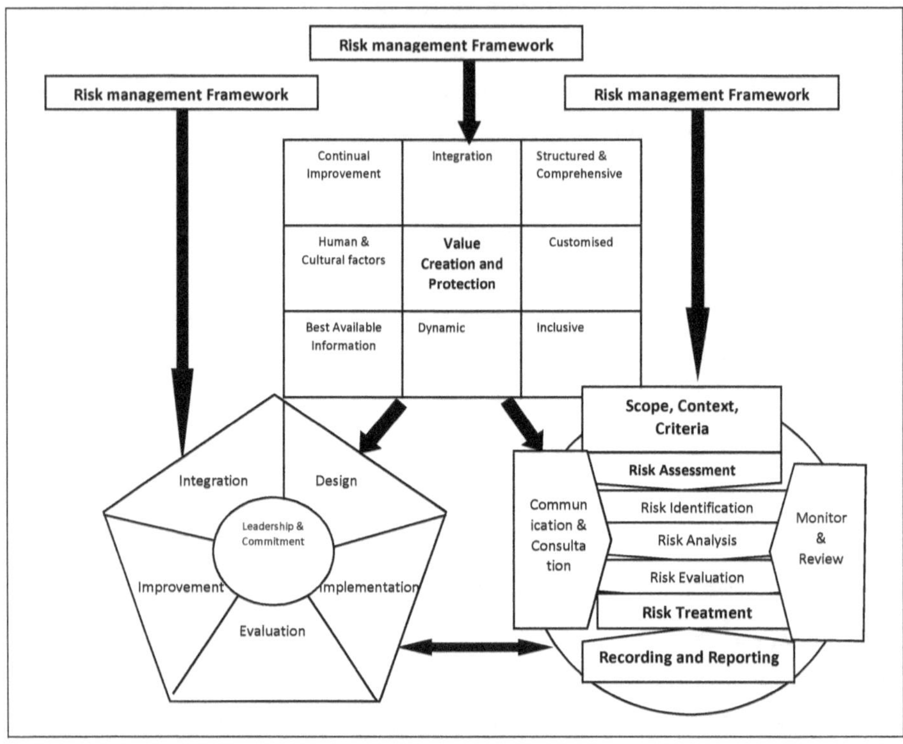

Figure 2.1: Principles, Framework and Risk Management Process.
Source: (ISO, 2009)

There is a link between the risk management principles, risk management framework risk management process presented in Figure 2.1. Risk management principles comprise components such as integrated, structured and comprehensive, customised, Inclusive, dynamic, Best available information, Human and cultural factors and Continual improvement. Risk management process includes scope, context and criteria; Risk assessment and risk treatment with the risk management framework being integration, design, implementation, evaluation and improvement (ISO, 2009). There are common phases within the risk management principles, processes and framework such as integration, evaluation, improvement and other facets which have also been utilised in this study.

19

Studies previously conducted in the previous four decades on risk management suggest that construction projects are exposed to risk at the time of their coming into existence and are perceived to have more inherent risk due to the involvement of many stakeholders such as owners, contractors and designers (Schieg, 2006; El-Sayegh, 2008; Forbes *et al.*, 2008). Traditionally, contractors in construction projects use high mark-ups to cover for risks but their margins have become smaller because this approach is no longer effective (Baloi and Price, 2003). Various stakeholders to a construction project have dissimilar behaviours facing project risk and different possibilities to transfer them to a third party that is in a position to deal with the risk (Kartam and Kartam, 2001).

The current systems used for risk management on construction projects focus on quantitative risk analysis although these techniques do not allow that risks, problems, remedial measures and lessons learnt from previous projects be captured and reused when developing new projects (Tah and Carr, 2001). Studies conducted previously concluded that over time and across countries, the construction sector tends to use only a limited number of techniques even though not all techniques are appropriate for every situation (Forbes *et al.*, 2008). One of the risk identification techniques for risk management includes brainstorming while qualitative methods of risk assessment are used most frequently compared to quantitative and semi-quantitative methods (Lyons and Skitmore, 2004). In addition, there are risk management techniques which are decomposition, artificial intelligence, probabilistic analysis, sensitivity analysis and decision making trees among others (Forbes *et al.*, 2008).

In view of the focus upon which the study is premised in terms of project constraints, construction project risk severely constrains the primary objectives of time, cost and quality which translate into additional cost and a lower return on investment to the client as well as a loss of revenue for the contractor (Visser and Joubert, 2008). It is also asserted that communication of construction project risk is poor, incomplete and inconsistent throughout the construction supply chain as the project participants also have a shared vision and understanding of the project. This subsequently leads to their inability to effectively implement early warning measures and mitigate strategies to adequately deal with problems resulting from decisions taken elsewhere in the chain (Tah and Carr, 2001).

It can therefore be concluded that the management of information and knowledge of a construction project are an essential part of a successful project risk management initiative because the knowledge management approach is a useful framework to improve the deficiencies of risk management process. Given the impact of an inadequate project risk management process in the final project performance, it is important to note that the development of risk management processes on construction projects are intended to identify its weaknesses and propose actions to reduce the identified weaknesses (Serpella *et al.*, 2014).

Risks in a construction project are mainly derived from two sources which are categorized as, on the one hand, environmental, also known as external risks, made up of those risks which are outside of the project influenced by issues outside of the project, societal risks, also referred to as Internal risks. On the other hand, pertinent to uncertainties existing within the project itself (Zhi, 1995). Some scholars also proposed a model which was referred to as the 'hierarchical risk breakdown structure' which is said to provide a basis for classifying risks within a project (Tah and Carr, 2001). The hierarchical risk breakdown structure allows risks to be separated into those related to the management of internal resources and the ones prevalent in the external environment as depicted in Figure 2.2. In the current study was confined and adapted to some of the internal risk sources of the hierarchical risk management breakdown structure and exclude the external sources in line with the essence of the study. Figure 2.2 provides the basis for classifying risks within a project and shows the hierarchical risk breakdown structure which places risks into two categories namely internal risks and external risks. The former describes local risk to be associated with issues such as labour, plant, sub-contractors, material and site dynamics as well as global risk which entails, construction, design, financial, location, pre-contract, client, contractual, environmental, financial institute, management and timeframe. The latter describes economic, physical, political and technological change (Tah and Carr, 2001). This study focused on the internal category of risks because the study needed to focus on the limitations expressed in the first chapter such as lack of skills, lack of experience and lack of finance. External risks were excluded from the current discourse the discussion would amount to a subjective task beyond the ability of this study. It is for this reason that the current study was viewed from a broad perspective of internal risks and cascaded down to specific risk areas from the numerous risks highlighted.

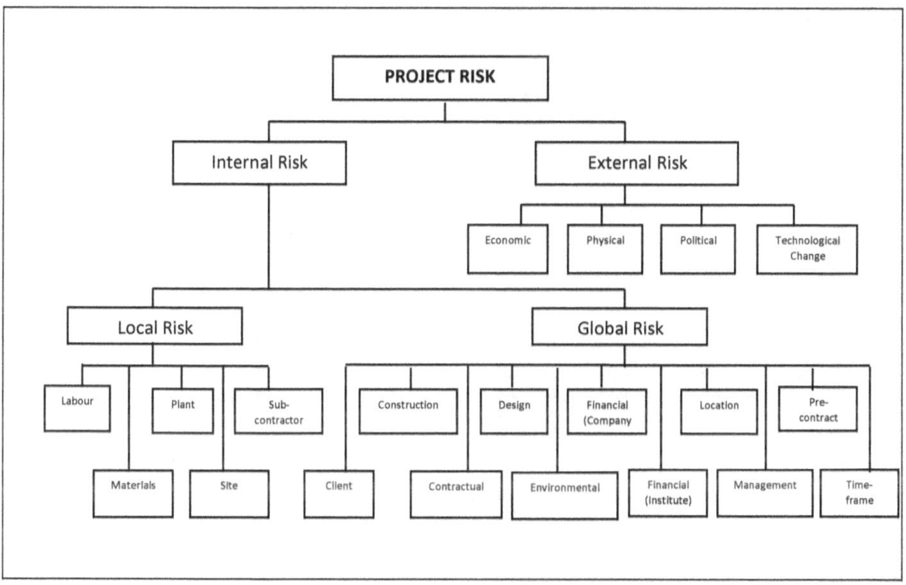

Figure 2.2: The hierarchical risk breakdown structure
Source: (Tah and Carr, 2001)

As stated previously in the preceding paragraphs, every construction project attracts risk in one way or other such that risk management becomes complicated and crucial for projects. During the pre-construction stage, there are numerous uncertainties that are considered and managing these risks at all stages of the project becomes important (Zhi, 1995). It is suggested that this view leads to a common and systematic approach whose distinct four stages are, risk classification, risk identification, risk assessment and risk response which this study will not delve into but only noted for the benefit of the readers' information. Identification will be briefly discussed and illustrated in Figure 2.3 below which depicts a list of common factors in line with the four risk classification groups;

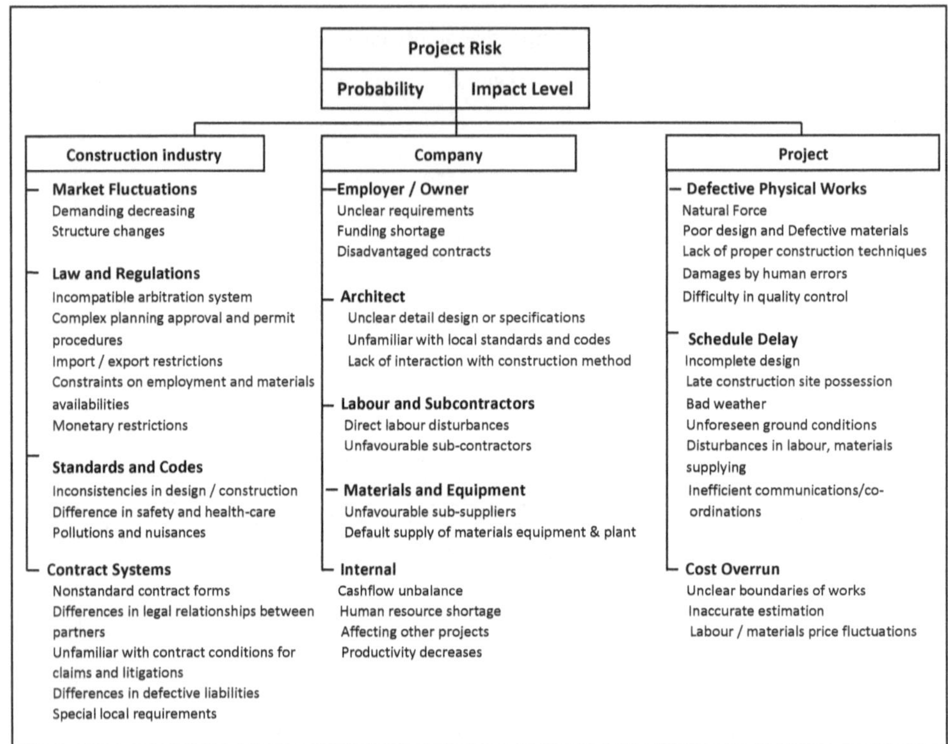

Figure 2.3: Risk Identification hierarchy for construction projects
[Adapted from (Zhi, 1995)]

Although risk factors differ from project to project, there are some factors which affect projects in general. The major risk factors for construction projects as depicted in Figure 2.3 at national or regional level relate to economic, political and social change while others are related to the construction industry, company and project (Zhi, 1995). The above assertion is supported by the view presented in the Hierarchical Risk Breakdown Structure which further categorises risks as, internal and external respectively (Tah and Carr, 2001). As indicated earlier, this study only focused on the internal risk factors and in accordance with the Risk hierarchy for construction projects presented in Figure 2.3, only the construction industry, company and project will be used as points of reference in this study. Figure 2.3 also presents internal risks inherent to the construction project which is depicted as company in Figure 2.3 under the internal category which highlights human resource shortages as one of the risk factors.

One of the many challenges faced by the sector regarding human resource shortages is the high turnover of staff within construction project circles.

2.2.1 The Effects of Risk Management on Project Constraints

In view of the fact that risk has either negative or positive effects on project objectives when it occurs, project success or failure is to a large extent dependent on the risk management strategy employed to mitigate that risk (Hillson and Simon, 2007; PMI, 2015). This therefore suggests that project success is also dependent on the way risk affects the project constraints of time, cost and quality is managed. Sound management of risk is a crucial determinant of the success of a project due to an increased attention to the variations in actual time, cost and quality compared to the expected ones as a consequence of a growing pressure on the reduction of time and costs (Cagliano *et al.*, 2015).

It has also been demonstrated through previous research that failure to deal with risk is one of the main causes of exceeding budget, falling behind schedules and missing performance targets (Carbone and Tippett, 2004). Other studies have further concluded that the negative effects of risk resulting from failure to manage it (risk), especially in the construction sector is exacerbated because projects are characterised by huge investments, long execution processes, many resources, many stakeholders as well as unstable economic and political environments which introduce a high level of complexity (Guofeng *et al.*, 2011). There is therefore a strong need to assess and control risks throughout all phases of the project which implies a multitude of methods which requiring the identification of circumstances under which each of them should be adapted and the criteria for choosing the risk techniques considered (Cagliano *et al.*, 2015). However, the set criteria do not take into account either a comprehensive set of unique characteristics of a project or its surrounding environment hence extreme caution must be taken when considering an appropriate risk technique to be applied.

2.2.2 The Time, Cost and Quality Triangle

The successful completion of a project requires several years of development, implementation, and evaluation of a successful project management strategy (Sutterfield *et al.*, 2006). This suggests that project success is be defined in many ways, one of which is to incorporate a wide variety of measures such as implementation within time, budget, performance and acceptance to the end user. Research previously undertaken revealed that since 1969, project success has been measured by the use of the iron triangle premised on the project constraints of time, cost and quality (Atkinson, 1999). An example of the Iron triangle is presented at Figure 2.4 below adapted from (Atkinson, 1999).

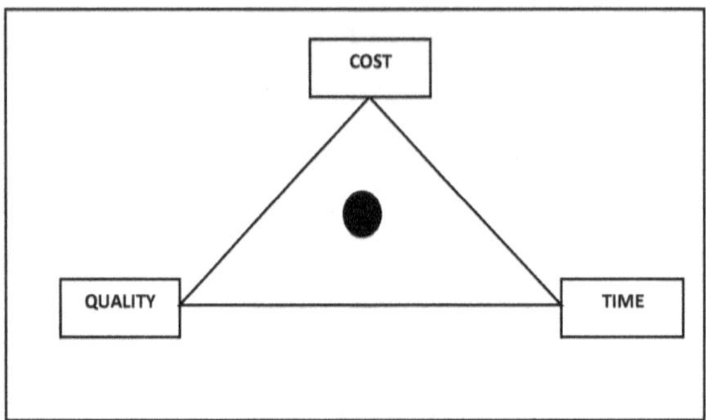

Figure 2.4: The Iron Triangle
(Atkinson, 1999)

The time, cost and quality triangle is mostly used in value management workshops as a tool to elicit from the client, their value criteria where the facilitor asks for team consensus on the position of the dot as presented in Figure 2.4 to indicate the teams' relative value criteria in relation to the three variables of time cost and quality. In this study, the three variables of time, cost and quality provide the basis of this research as criterea set for project success. Scholars further concluded that the identification of other factors which have been revealed as definition of stakeholder needs, expectations, safety, efficient use of resources, effectiveness, satisfaction of stakeholders, reduced conflicts and disputes and project tasks is critical to project success. However, the current study was confined to the project success criterea of time, cost and quality despite the fact that some projects still fail even when the time, cost and quality triangle is in place. This is because in certain cases, failure is still recorded in many projects as the term failure is normally used when a project is terminated prior to completion which might suggest that such failure is caused by other factors such as legal, social, political, technological and environment reasons (Pinto and Mantel, 1990).

The effects of the constraints in Figure 2.4 against each other are that time, which entails any delay in the schedule affects cost and this has a tendency of escalating over the passage of time regarding increase in the cost of material as well as overheads such as wages and plant in view of standing time. Cost when not properly managed has a two dimensional effect on projects as it affects quality and scope. This is because when material is not procured on time and prices escalate during the period, the quantities of material procured is less and this. The delay may lead to a compromise on quality and less material is used for larger proportions as well as budget overrun where material is applied strictly in accordance with the specifications. An increase in cost affects the overall budget where fewer materials are procured thereby affecting the expected project delivery time. In certain cases, quality is compromised as the increased cost causes fewer materials to be bought which in-turn are stretched to either meet or accommodate the scope of work thereby compromising quality in the process. Finally, quality is a constraint which affects the workmanship of as poor workmanship causes poor quality output and it results in re-works. Every re-worked activity entails usage of new material which is procured at an extra cost. The re-work also has an effect on the schedule because the time spent on corrections or re-working poor quality work has an effect on the overall project schedule.

2.2.3   *Risk Management Tools and Techniques*

A description of the risk management tools currently in use within the construction industry are outlined in brief in order to provide the reader with a background of the tools and techniques in use although this study did not present a detailed discussion on the same. The findings of research previously conducted identified some of the risk management techniques currently in use as, Risk Premium, Risk Adjusted Discount Rate, Subjective Probability, Decision Analysis, Sensitivity Analysis, Monte Carlo Simulation, Stochastic Dominance as well as the Casper and Intuition (Akintoye and MacLeod, 1996). In addition, these researchers also identified some of the risk management tools currently in use as Algorithms, Mean and Analysis, Bayesian Theory and Decision Trees whereas the strategies for risk allocation were identified as Retention, Reduction, Transfer and Avoidance (Akintoye and MacLeod, 1997).

In addition to the above, it is important to note that the risk management process consists of processes which require different techniques to manage such risks for every respective process and the techniques are, Risk identification, Risk assessment, Risk response and Risk control (Mahendra *et al.*, 2013). These techniques which are within the respective main processes are, Risk Identification (Past experience and Checklist), Risk Assessment (Sensitivity analysis, Scenario Analysis, Probabilistic Analysis, Monte Carlo Simulation and

Decision trees). Other techniques are, Risk Response Planning (Risk exploit, Risk share, risk enhance, Risk acceptance, Float and Contingency), Risk Control: Other techniques are identified as Program Evaluation and Review Technique (PERT), probability distributions, Risk Breakdown Structures, Fuzzy Sets Theories (FST), and the Probability Impact Grid (P-I) Model.

2.3   VALUE MANAGEMENT

Value management emphasises effective and cognitive participative decision making processes for setting and implementing specific values and specific project goals (Leung and Liu, 2003). Value management is therefore defined as a systematic problem solving approach to achieve the desired quality at the lowest cost (Leung and Liu, 2003). This definition resonates with the purpose of the current study where the application of the Integrated Approach for Value management is being tested as a risk management tool.

Traditional value management processes are undertaken in three different stages over a period of time and these are, the pre-study stage, the value study stage and the post value stage (Leung and Liu, 2003). The stages are similar to those undertaken during the tenure of this research. A typical Value Management Process is depicted in Figure 2.5 below:

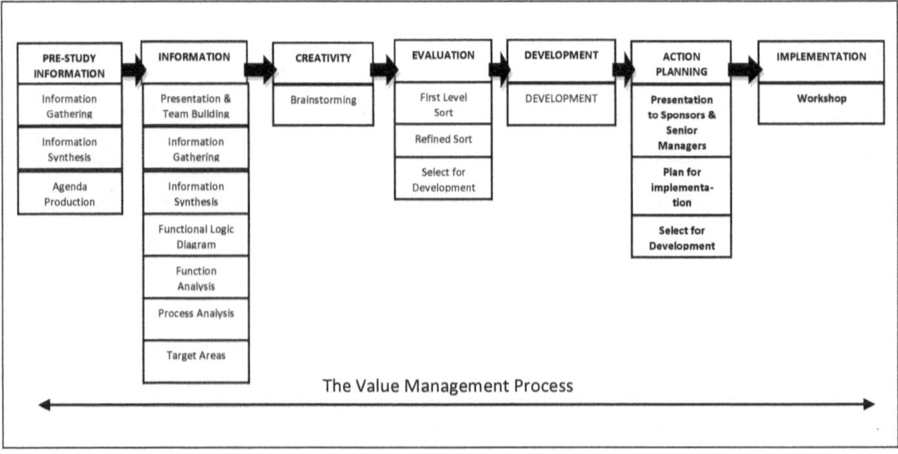

Figure 2.5: The Generic Value Management Process

(Kelly *et al.*, 1998)

The generic value management process presented in Figure 2.5 above, reflects a basis for the benefits it provides in construction as the aim of employing this process is to achieve maximum effect on any project during its lifecycle (Kelly *et al.*, 1998). This study draws interest in the Generic Value Management Process because of the drivers upon which the process is premised such as the pre-study information, information, creativity, evaluation, development, action planning, workshop report and implementation. These drivers resonate with the processes adopted for this study adapted from the phases within the action research cycle in which only those drivers which the researcher employed for this study are emphasised. This process finds prominence in this study because it acknowledges the importance of using existing project teams as a base for the selection of a value management workshop team which brings experts to the workshop. The advantages are; reduction of cost, pool for better ideas due to greater experience 'second' time around, offers another chance to explore alternative options, useful for teambuilding and resolving conflicts that may exist within the team, increased communication and increased implementation.

2.3.1 *The International Benchmarking of Value Management*

This was done to ensure that the findings at every stage of the process inform or determine the type of process to undertake during the next stage. Scholars who previously wrote on the subject believe that planning as an activity is conducted during the pre-study stage where the status of resources is evaluated and project information is analysed in order to address the information phase of a value study. This is followed by the second stage which is comprised of six phases which are, the Information phase, Functional analysis phase, Creativity phase, Evaluation phase, and Recommendation Phase (Leung and Liu, 2003).

According to the post-study stage involves the implementation, verification and feedback processes. However, there are other researchers who mainly concerned themselves with value management in terms of its technical processes such as brainstorming, Functional Analysis Systems Technique (FAST) diagrams, Tree diagrams, weight evaluations and cost reductions in life-cycle costing in order to systematically solve the "hard" technical problems such as cost savings (Green, 1999a). This approach led to the re-definition of the value management concept as stemming from a "System theory". The primary objective of Value management is for the project participants to develop a common understanding of the design problem, to identify explicitly the design objectives and to synthesise a group consensus about comparative methods and alternative courses of action.

This current research opted to adopt the initial definition of Value management depicted in paragraph 2 of Section 2.3 above. While other researchers have used the terms Value engineering (VE), Value Analysis (VA), Value Methodology and Value management interchangeably to refer to the same concept. In this study the researcher adopted the use of Value management as the preferred term for this purpose because this term resonates with the subject and purpose of this study and also distinguishes itself from other cost reduction activities (Haas and Hansen, 2005). It is also believed that Value management has become a proactive problem solving process which is used to enhance the functional value of a project by managing its development from design concept to operational use and decommissioning (Hayles *et al.*, 2010).

The findings of some studies indicate that the value management approach aims to exploit the synergistic benefits obtained from gathering relevant project stakeholders together as a group. This is achieved through structured, team oriented and open-dialogue exercises which recommend alternatives or confirms existing solutions by referring to the value requirements of the client (Bowen, Jay, *et al.*, 2010; Hayles *et al.*, 2010). Researchers who conducted studies on the awareness of the application of Value management among professional construction managers and construction project managers in South Africa revealed that the awareness of value management was not widespread among these construction professionals and its impact on construction projects was minimal (Bowen, Edwards, *et al.*, 2010). The finding suggests that while there is so much value management information available to construction professionals in South Africa, Value management techniques have not been utilised to full potential and the intended effects on cost reduction, time and quality have not been achieved which also validates the basis of this current research.

Value management is also described by other scholars as a structured and analytical process aimed at achieving value for money by providing necessary functions in projects at the lowest cost consistent with required standards of quality and functionality (Standard, 2007). The concept is further defined as the name given to a process in which the functional benefits of a project are made explicit and appraised consistent with a value system determined by the client (Kelly *et al.*, 2014). While Value management has its origins in the manufacturing industry, its application in the construction sector has been a subject of considerable research which inspired the basis upon which its theoretical framework is premised (Kelly *et al.*, 2014). Accordingly, some of the research upon which the framework or structure of value management is based are: Advocating the use of value management in construction (Dell'Isola, 1982; Connaughton and Green, 1996), best practice value management and benchmarking (Male and Kelly, 1998); value management for managing the project briefing and design processes (Yu, 2002; Kelly *et al.*, 2014), as well as the integration of risk and value management (Green, 2001; Dallas, 2008). Some researchers argue that the development of an underlying framework for value management is conducted through an in-depth study of relevant literature from several fields such as general management, purchasing, operations, marketing, quality and logistics (Dumond, 2000).

The literature from the sources revised for the current study leads to the clustering of related concepts to render a reasonable but accurate theory which is analysed. Further, the concepts cause the creation of the ten primary components, which are an important contribution to the structured framework of value management but will not be discussed in detail (Dumond, 2000). The ten primary components referred to are presented as follows:

i)   Project mission; Initial clearly defined goals and general directions.

ii) Top management support; willingness of top management to provide resources.

iii) Project schedule or plan; A detailed specification of the individual action steps.

iv) Client consultation; Communication, consultation and active listening to all parties.

v)   Personnel; Recruitment, selection and training of necessary personnel.

vi) Technical tasks; Availability of the required technology and expertise to accomplish it.

vii) Client Acceptance; The act of selling the final project to its ultimate intended users.

viii) Monitoring and Feedback; Timely provision of comprehensive control information at each stage of the process.

ix) Communication; The provision of an appropriate network and necessary data to all key actors in the project implementation.

x) Troubleshooting; Ability to handle unexpected crises and deviations from plan.

(Pinto and Slevin, 1988; Dumond, 2000)

The relations diagram presented in Figure 2.6 below depicts the associated relationships in the development of a theory and clarifies the conceptualisation process (Whetten, 1989).

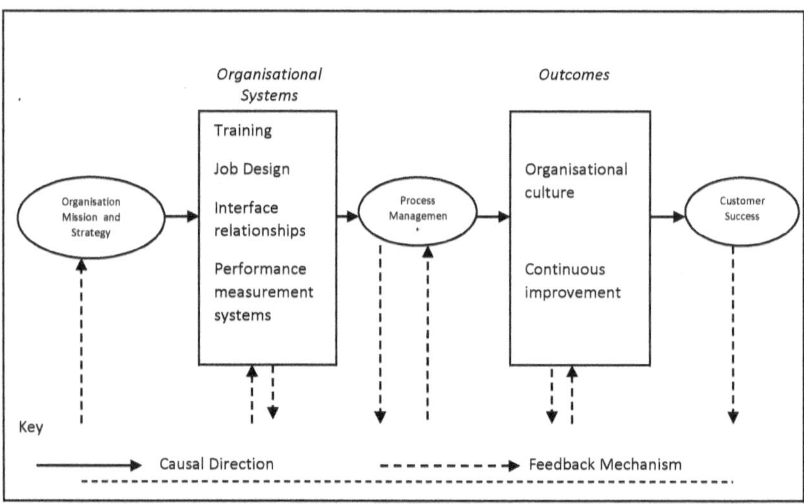

Figure 2.6: Proposed framework for value management
Adapted from (Whetten, 1989)

As depicted in Figure 2.6, the framework for value management illustrates that management drives value management through the mission as well as strategy and subsequently, the organisational system that is developed for the organisation. The linkages among components are presented through a relationships diagram which shows causal relationships in the development of theory and permits the researcher to clarify the conceptualisation process when many complex variables are present (Whetten, 1989). Similar to the current study, the illustration in Figure 2.6 is typical of the project upon which this study is based. In setting the project upon which this research is based, the project, represented by the organisation, mission and strategy component in the first component of Figure 2.6, has a mandate to achieve project success based on the criteria set which is presented as the organisational system in the proposed framework for value management. process management in this study is represented by the action research undertaken while the outcomes are denoted by the operational changes in the project resulting from the stimulating actions introduced to project activities. Customer success in Figure 2.6 is represented by the successful delivery of the project in accordance within the set criteria of time, cost and quality.

The value management framework is based on the fact that pressure for increasing quality while reducing time and cost places particular emphasis on managing risk in projects (Cagliano *et al.*, 2015). The framework referred to in the preceding discussion provides guidelines for the selection of risk techniques with due regard to most relevant aspects characterising the managerial and operational scenario of a project. The theoretical framework to classify these techniques is the phase of risk management process, the phase of life-cycle and the corporate maturity towards risk which in turn enhances value (Cagliano *et al.*, 2015).

2.3.2 *Sound Management of Risks*

Sound management of risks is a crucial determinant of project success in view of its increased attention to the variations in actual quality, time and cost performance compared to the expected ones as a consequence of a growing pressure on reducing time and costs as it has been demonstrated that failure to deal with risk is one main cause for exceeding budgets as well as falling behind schedules and missing performance targets which result in the need for a candid framework for Value management (Carbone and Tippett, 2004). While risk management processes and supporting techniques have been extensively developed and implemented, the multitudes of methods demands for instruments suggesting under what circumstances each one of them should be adapted as these criteria usually do not take into account either a comprehensive set of the unique characteristics of a project and its surrounding environment hence the development of a theoretical framework of Value management. The 'Risk influence matrix' and the 'Aliens eyes' have been identified as risk models while the risk mitigation frameworks have been identified as the 'quantitative risk mitigation framework' (Wang *et al.*, 2004). Therefore, the above implied that it is important to employ both matrices in order to attain more value from the risk management tools used.

2.3.3 *Value Management Tools*

Tools used to manage risks within the value management domain are, the Functional Analysis Systems Technique (FAST), Simple Multi-Attribute Rating Technique (SMART), Kano Model, Lever of Value, Quality Function Deployment Technique (QFDT), REDRESS, Spatial Adjacent Programming (SAP), Time Cost and Quality Triangle, Value Analysis and Life-Cycle Costing (Bowen, Jay, *et al.*, 2010; Hayles *et al.*, 2010; Bowen *et al.*, 2011). In view of the detailed and extensive literature on each of these tools, this current research only confined itself to the discussion of the Time, cost and quality triangle being the area within which this research is focused.

2.3.4 *Phases within the Value Management Cycle*

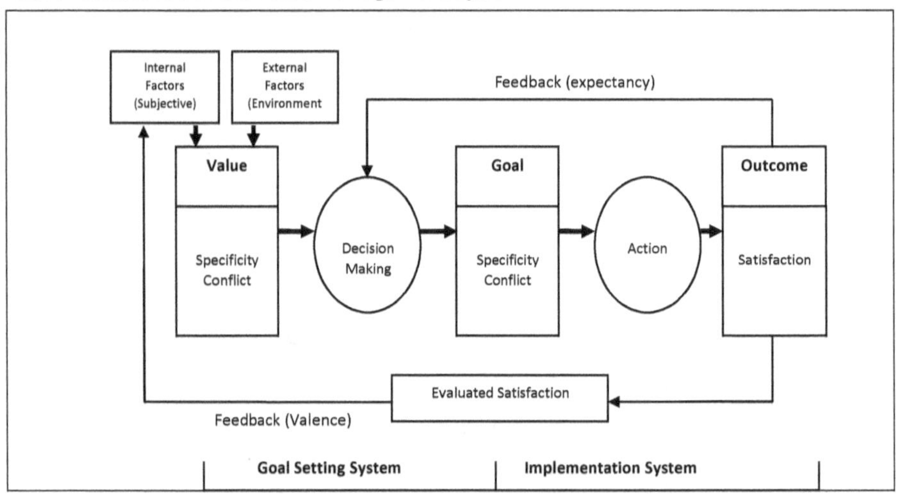

Figure 2.7: A Conceptual Model of Value Management
SOURCE: (Leung and Liu, 2003)

Similar to the current study, the illustration in Figure 2.7 is typical of the construction project upon which this study was based. In the setting of the construction site upon which this research was based, the project, represented by the organisation, mission and strategy component of Figure 1, has a mandate to achieve project success based on the criteria set which is presented as the organisational system in the proposed framework for Value management. Process management in this study is represented by the action research undertaken while the outcomes are denoted by the operational changes in the project resulting from the stimulating actions introduced to project activities. Customer success in Figure 2.7 represented by the successful delivery of the project in accordance with the set criterea of time, cost and quality

2.4   PROJECT SUCCESS CRITERIA

In order to determine whether a project is successful or not, the criterion set prior to the commencement has to be used as a yardstick of this measurement. Therefore, success criteria are measures against which the success or failure of a project is judged (Cooke, 2002). For the purpose of this research, project success criteria are defined as the set of principles or standards by which favourable outcomes can be completed within a set specification (Chan and Chan, 2004). Owing to the fact that project success is envisaged for every project, project functionaries especially the project managers ensure that all resources needed to meet the success criteria are mobilised and applied to the project because success criteria for project management are often measured on the basis of time, cost and quality. The constraints of time, cost and quality upon which basis the success criterea is measured has proved to be limited and suggestions in terms of the criterion of time, cost and quality outlined by authors in previous discussions, it is impossible to generate a universal checklist for project success criteria suitable for all projects (Westerveld, 2003).

Research previously conducted by Wateridge (1998) suggests that a success criterion has no consistent interpretation or methodology for measuring it although the success criteria must be set out clearly and agreed upon at project inception to avoid conflict and differences amongst project teams. All stakeholders need to contribute to the development and approval of the criteria in order for the project to concentrate on the agreed success criteria and factors which translate into results (Wateridge, 1998).

When project objectives are clearly established and challenges outlined, project teams are able to use their endeavours to deliver the project in accordance with the objectives within the constraints (Cooke, 2002). One of the many challenges arising resulting from the inability to establish clear project objectives is identifying the stakeholders who should decide the criterea because until there is an agreement on the determinants of success, it would be difficult and almost futile to accurately monitor and predict project results effectively. This is the basis upon which project success plays a critical role especially on construction projects (Atkinson, 1999).

Criterea can be consistently reviewed with project progress as they are subjective given the dynamic nature of project risk sources which are external and internal (Wateridge, 1998). The subjective measures include quality, functionality, end user satisfaction, design team satisfaction, client satisfaction and construction team satisfaction. Objective measures include environmental impact assessment scores, speed of construction, variation over final cost, construction time, unit costs, net present value, accident rate and time variation (Chan and Chan, 2004). However, other researchers concluded that a project can fail to meet the cost and time expectations but still be considered a success (Hough and Morris, 1987; Chan and Chan, 2004). There are a few examples cited to qualify the claim that a project may still be regarded as successful even when it is completed beyond schedule and at a cost over-run such as the Sydney Opera House which is currently considered an engineering masterpiece yet it took 15 years to build instead of the scheduled five years. The project and was 14 times over budget as did the project of the barrier on the River Thames in London which took twice the duration to construct at a 400% cost overrun yet it is considered a success from the perspective of both the client, the users and the contractors, who construed it to be profitable (Wateridge, 1998; Jugdev and Müller, 2005). The failure of the Sydney Opera House project was construed in the context of project success while its success was based on the set project success criteria. Similarly, the River Thames barrier project was perceived as a failure in the context of time and cost although construed to have been a success on account of satisfaction derived from the use of the project as a product by some influential sectors of stakeholders as well as the contractor who based the project success on profitability.

Another perspective to the examples of the Sydney opera house and the River Thames project reveals that where stakeholder management has not been effectively carried out, the end product is mostly one which is not in accordance with the requirements of the customer or user clients. There are three reference points in terms of this claim in which the multi-million rand Bophelong taxi rank project in Vanderbijlpark, South Africa has become a white elephant after its perceived successful completion. This is because the taxi owners could not operate in the newly constructed taxi rank as a result of the height of the structure that mini-bus taxis could not drive through. (Reid, 2006; Beavon, 2001). In addition, the construction of gantries for e-tolls on the Gauteng highways and the construction of the Cape Town Stadium are two other projects which were embarked upon without a proper and effective consultative or stakeholder management process (Chain, 2009; Preuss *et al.*, 2014). In terms of the e-tolls, motorists, being one of the most influential stakeholders in their capacity as the end-user client, should have been engaged in order to get their views about the project before its implementation. This would have prevented what has become of the project where motorists refuse to pay e-tolls rendering this a failed project (Musakwa, 2014; Goetz and Schaeffler, 2015). An association called the Association Undoing Tax Abuse (OUTA) representing the interests of the motorists has been opposed to the implementation of the e-tolls in a vocal manner (Naidoo, 2013). The construction of the Cape Town Stadium as part of the 2010 world cup legacy project suffered the same fate due to lack of an effective consultative process with influential stakeholders which has left the stadium underutilised contrary to its intended purpose (Chain, 2009; Preuss *et al.*, 2014).

2.5   INTEGRATION OF RISK MANAGEMENT AND VALUE MANAGEMENT

Risk management and Value management are both well-established disciplines recognised for their part in best practice as the two disciplines are both compatible and complementary which presents them with a great potential for a common framework (Hiley and Paliokostas, 2001). Researchers concluded that risk management enhances value management through the use of the value management team to conduct the risk management plan of the project especially during the creative stage of the project when alternative methods of mitigating recognised risks are generated (Norton and McElligott, 1995). While a combined approach affords the project potential benefits derived from the assembled multi-disciplinary team, which also promotes the introduction of risk management into the project, the combined approach has an advantage for a creative and positive value management atmosphere required for the generation of ideas to mitigate risk (Connaughton and Green, 1996; Thiry, 1997).

Two processes which involve stakeholders are identified as 'Soft Value Management' and 'Hard Value management' in which the former is considered to occur very early in the process when the project objectives are not yet defined while the latter is a process initiated when design information is available (Connaughton and Green, 1996). The fact that value management is concerned with resolving the uncertainty associated with project objectives while risk management addresses the uncertainty associated with outcomes means when risk management and value management are integrated, they are expressed in terms of uncertainty where the interdependence between them becomes apparent (Green, 1994). In a further attempt to address the integration of value management and risk management, scholars suggest that a generic approach to value management which is encompassed within the traditional risk management programme should also include a qualitative risk assessment, a quantitative risk analysis and also the implementation of risk mitigations (Kirk, 1995). Scholars suggest that the integration of the two disciplines provides the means to for a more effective use of the team which translates into the effectiveness of risk responses to enhance the use of value management which in turn increases the project value (Hiley and Paliokostas, 2001).

It is also worth noting that the creative atmosphere of value management identifies solutions for specific risks that the use of risk management alone misses hence the satisfaction derived from the integrated use of value management and risk management which leads to a complete picture of the project. While the 'pros' of employing the integrated approach outweighs the 'cons', it is important for the purpose of this study to briefly outline the disadvantages of using the integrated approach of the two disciplines. Risk management and value management are considered difficult to carryout adequately within severe time constraints as their integration worsens the problems associated with (Hiley and Paliokostas, 2001). The two disciplines have further not received the recognition they deserve in the construction sector and the danger emanating from the integration has the potential of rendering the techniques dominant with the corresponding loss in the other's benefit (Hiley and Paliokostas, 2001). It is therefore concluded that while the integration of both disciplines enhances the outcomes and improves the efficiency, which also leads to credibility, risk management adds a further dimension to the evaluation of value management proposals. The integration also increases the knowledge of a project which results in improved decision making.

2.5.1 The Integrated Approach for Value Management

Solving problems which impede project success requires the application of a multi-disciplinary approach in order to retain an acceptable level of understanding of project objectives. Research previously conducted by Al-Yami (2008) concluded that Soft Value Management (SVM) is a reliable tool used at strategic level for producing visions of a new direction and filtering objectives. According to the views of Al-Yami (2008), Soft situations are classified as those which are not well defined and cannot be described within a clearly defined system boundary while 'hard' situations are those which may be described within a clearly defined system boundary. Therefore, SVM is used to define and describe clearly defined issues to help its implementation and the integrated approach which is uses six processes adapted from value management. The main stages of SVM are outlined in Figure 2.8 below (Barton, 2000).

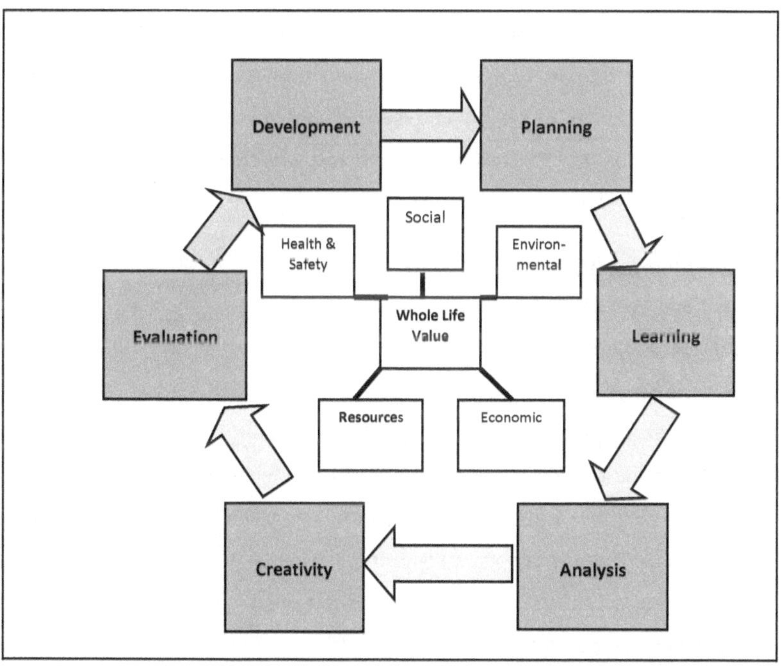

Figure 2.8: An Integrated Approach for SVM
Source: (Barton, 2000)

This study adopted a mixed methods approach in which the deductive component entails the development of a theoretical framework prior to its testing through empirical observations, which in the case of this current study is done through action research in accordance with Barton (2000) as well as Gill and Johnson (2010). The Integrated approach presented in Figure 2.8 is synthesised through a review of the related literature and best practice and reinforced with information filtered from interactions with project stakeholders who possess significant experience in the construction sector with a specific bias in the areas upon which this study is focused. Soft Value management was developed through action-research to determine client needs and requirements during the different phases of the project and the process of capturing requirements involves identifying the values of stakeholders to take into account its implementation project in which Value Management through action research is described as a structured facilitated, process in which decision makers, stakeholders, technical specialists and others work collaboratively to bring about value based outcomes in systems, process, products and services (Barton, 2000; Al-Yami, 2008). The processes forming part of this study were adapted from the views advanced by scholars in the preceding paragraph regarding the attainment of value management through action research in which all project stakeholders and role players work corroboratively to bring about value based outcomes.

2.5.2 Synthesis of Risk and Value Management

Risk Management and Value Management are widely used in project management as best practice to facilitate successful project delivery (Hiley and Paliokostas, 2001). Effective utilisation of risk and value management methodologies aid project managers to reduces uncertainty and risk exposure while maximising value and return on investment (Green, 1999a; Dallas, 2008).   Due to the fundamental similarities between RM and VM processes, scholars and practitioners have been advocating for the integration of the two in a single process (Green, 1999a). These proponents argue that having an Integrated Risk and Value Management (IRVM) study process delivers better project outcomes with enhanced Value for Money (VfM) when compared to independent practice. Managing risk in the short and long terms is an integral component of achieving any value proposition therefore both risk and value should be managed concurrently. Further, considering the significance of similarities between risk management and value management, the duplication of effort in handling the two processes is eliminated.

The use of an integrated approach for both disciplines has also found prominence in the Public Private Partnerships (PPP) for project management. Scholars say that the integration of risk management and value management in Public Private Partnership (PPP) project, management establishes a clear synergy between minimising risk and maximising value in a project or program and researchers and practitioners further argue that integration of risk and value management delivers better outcomes to the client than separate application of the two processes (Akintoye, 1996). Accordingly, this research was aimed at exploring the potential benefits of integrating RM and VM in a single study process along with the identification of the Project Success Criteria (PSC) related to the successful implementation of an Integrated Risk and Value Management (IRVM) process. As a means of identifying the approaches used in combining the two processes, this study put forward a case study for the application of an integrated risk and value management study through action-research where it is conducted in accordance with and reflects the standards of, Value Management adapted from the models presented in the preceding sections.

2.5.3 *Management approaches for current construction projects*

The public procurement system of capital infrastructure projects in South Africa is managed through the National Department of Public Works, which essentially prescribes procurement processes which are mainly done through the bid system, by means of 'closed' and 'open' tenders (Kaseke, 2011). This implies a procurement system in which a tender is advertised in public and specifications for compliance are prescribed where a successful tender is awarded from the complying submitted bids (May and Govender, 1998). It therefore suggests that procurement by open bid is carried out for tender classified as 90/10 procurement relates to tenders of a value exceeding R1 000 000 (R1m). In terms of closed tenders of bids below the R1 000 000 thresh-hold, also categorised as 80/20, three bidders are selected from the central database system of government and these bidders are invited to submit quotations based on the prescribed criteria and the bidder who best suits the criteria is awarded the project (Walker, 2003). The procurement method for sub-contractors on school construction projects is done through the close tender system by both the client and the successful bidder by way of nominated subcontractors and local subcontractors respectively (Bolton, 2007).

Other scholars are of the view that there are several risks which are faced by each respective stakeholder. In view of time constraints and for the benefit of the reader, this study will reflect the risks pertinent to only four stakeholder categories which are; the client or customer, the team, community and the contractor. The risks associated with stakeholder categories are summarised as, civil unrest, protest action, construction time, natural disasters, abandoning of the project by the contractor, poor quality work, uncertain productivity, weather or seasonal implications, financial risk such as inflation, wrong estimates, bureaucracy in the system, fluctuating exchange rates which are ascribed to the client (Skitmore, 2003). Some risks common to the professional team are stated as lack of coordination, constant scope changes by the client, lack of detail in the concept by the project brief, use of wrong tools and equipment, design problems and delays in project implementation (Walker, 2003; Bolton, 2007). Some scholars also noted some risks common to the community as, environmental (air, water and soil pollution), noise, health, injuries and delayed payments by the contractor and sub-contractors (Chow, 2007). Finally, in terms of potential risks to the contractor. Other scholars assert that safety hazards, managing change orders, incomplete drawings, design errors, poorly defined scope, unknown site conditions, poorly written contracts, unexpected increase in material costs, labour shortages, damage or theft to equipment and tools, inclement weather, sub-contractor problems, poor project management and non-availability of materials are some of the potential risks faced by the contractor (Walker, 2003; Bolton, 2007).

2.5.4 *Researchers conceptual model for VM/RM application*

It is argued that value for money can be achieved either by enhancing the requirements of a project or by reducing the cost of meeting them. Therefore, the search for value for money means an attempt to find the best balance between meeting the requirements of the stakeholders and the resources available (Connaughton and Green, 1996). Finding this best balance will inevitably involve some risks and these risks have to be identified and assessed. The SMART value management process favours shared views and perceptions rather than optimising views. It should be recognised however, that shared views and perceptions are themselves accompanied by uncertainty, hence there is no right answer but rather a consensus on the objectives. To try to resolve this uncertainty would not be pragmatic but the assessment of risk and risk perceptions can at least provide a valuable framework for controlling the value management outputs.

All the methodologies and approaches described in the preceding discussion refer to risk issues relevant to specific areas within the value methodology and in most of these cases, the focus is on project costs and on schedule. Furthermore, all these methodologies seem to be undergoing continuous improvement and there is no definite strategy at present amongst project management practitioners. However, they can definitely be seen as major stepping stones for exploring further the possibility of developing an integrated approach to risk and value management that will improve value and manage risk effectively for a project.

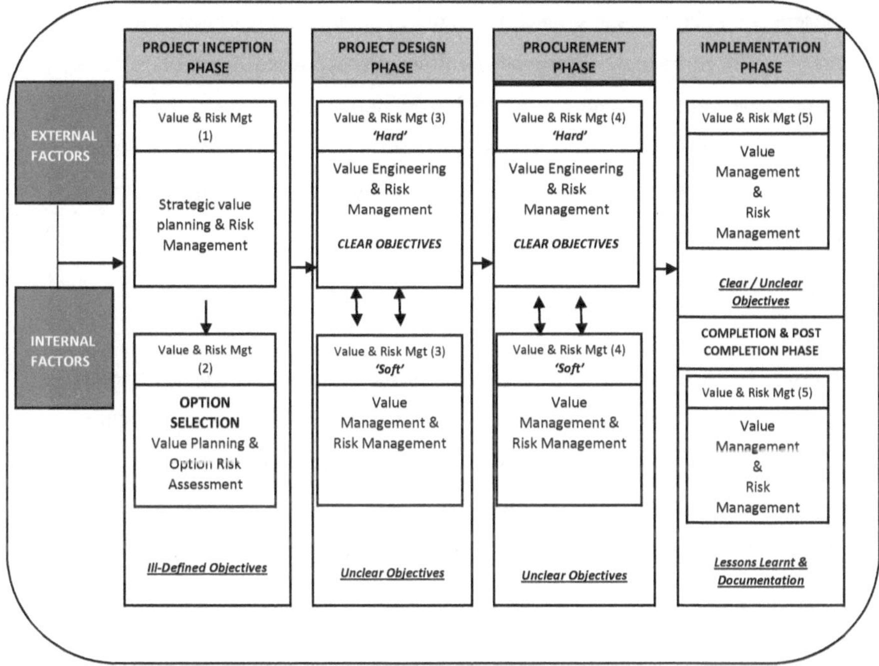

FIGURE 2.9: RESEARCHERS PROPOSED CONCEPTUAL MODEL FOR VM/RM APPLICATION
[Adapted from Mootanah (1998)]

The analogy of the conceptual framework in Figure 2.9 conceptualised the researcher's perceived model integrating risk management and value management. The possible outcomes of these relationships are that (i) the risk and value interface could be dynamically realisable by integrating value and risk methodologies as and when required in project management, depending on the type of problem situation at any point in the project's life; (ii) some partial interaction or integration could be possible with interventions at project management phases and (ii) full integration of risk and value management could be possible and relevant for specific projects, for specific problem situations at specific times in the project's life. A conceptual framework for integration was thus developed as adapted from Mootanah (1998) and presented in Figure 2.9.

Successive phases at which integrated value and risk management processes were carried out in the management of a typical construction project. These processes would have helped to achieve consensus on objectives at project definition, to determine the design options at the design phase, to select the most appropriate contract strategy at procurement, to solve problems during implementation and to record lessons learnt for future projects. The conceptual framework took into account the hard-soft nature of problem situations presented as Annexure H within a project as well as the project life cycle dimension in the application of integrated value and risk management.

2.6 SUMMARY

This chapter presented a comprehensive literature review undertaken in which various sources such as journal articles, magazines, books, and previous research were consulted. The process for this study commenced with a literature review on the topics of risk management, value management, project success criteria and the integration of risk management and value management. The chapter also evaluated the current risk and value management processes, frameworks, tools and techniques and also presents the views of different scholars on the subject of this study. Value engineering, value analysis, value planning, success criteria, whole life value, whole life cost and value for money are also explored. Each of these areas were covered in general and especially in terms of relationships between the risk management and value management subjects. Over 100 citations are reviewed to define the scope of this study and explore the conceptual linkages between value management and risk management in order to meet the aim and objectives of the research. Although the researcher experienced some challenges to acquire sources which directly dealt with some areas of this study, an attempt has been made to refer to literature which is closely related to the field of this study. It is also worth noting that while there are attempts in construction research towards finding a single framework for risk management, the concept of blending risk management principles into value management by identifying the conceptual linkages, underlying logic of well-established and integrated approach to risk management within project management strategies emerged as the thrust of this research.

The literature review for this research reflects the various techniques used for RM and the views of various writers on the definition, application, benefits and the methodology of value VM. It provides a base as a theoretical perspective within the value management field as a subject. The insights given by the literature also provide additional information upon which this study relies such as the different approaches to attaining value, identification of risk, definition of quality and risk management in general. A discussion on the current risk categories faced by each stakeholder on construction projects and also the procurement systems used to procure materials and services for projects is also presented. The analysis of the literature reviewed formed the basis and framework of the research book although some situations were adapted to the principles derived from the literature in order to provide a solid basis for the current study.

CHAPTER 3: RESEARCH METHODOLOGY

3.1 INTRODUCTION

The previous chapter presented a critical review of literature associated with all aspects of the field of this study where the use of value management for risk management was presented and also identified tools and techniques currently used for risk management on projects in general. The chapter further presented literature on framework, processes and integration of risk and value management and also explored the effect of risk on project success according to different views of supporters and critics of value management.

This chapter contains the 'Research Methodology' which is described as the mode of operation for the appraisal of research claims and justification of the knowledge gathered from literature (Creswell, 2013). A research methodology is a strategy used by a researcher to undertake a study (Gill and Johnson, 2010). This chapter also describes and explains the research methodology and strategies used as well as a theoretical justification of the choice of critical theory of epistemological paradigm and action-research as a methodological paradigm. Also presented in this chapter are methods of data collection, data analysis and research purpose, research philosophy, research methodology, research strategy and ethical considerations. The study was conducted within the context of reviewing risk management tools and the implementation of project management strategies. Table 3.1 depicts the methodological options adopted for this study.

3.2 RESEARCH PURPOSE

The subject of this study is the review of risk management tools for construction projects and explored the implementation of project management strategies. The purpose of this research was to review the extent to which value management was used for risk management and ultimately evaluate the implementation of project management strategies through action research on an existing school construction project. This was done by implementing value management techniques on project activities in order to elicit opinions of different scholars and validate the use of value management techniques to manage risk on projects. The methodological options adopted by the researcher for this study are presented in Table 3.1.

Paradigmatic Assumptions			
Epistemological Paradigm		Critical Theory	
Methodological paradigm		Action Research	
Selection Procedures			
Selection of Site		Purposive Selection: A school construction site where the role of the researcher is that of Project manager	
Data Generating Techniques		Data-documentation techniques	
Analysis of trends and activities Observations		Physical recording	
Data Analysis and Interpretation Thematic analysis and interpretation			
Quality Criteria of the Study			
Credibility Physical recording of pre and post action project status	**Process Validity** Rich description of research process	**Dependability** Document trail	**Catalytic validity** Evidence of how the stimuli introduced to project activities stimulated change action study stimulated
Ethical Considerations			
Permission was granted by the Ethics Committee (EiR) of the university for this study to be conducted			

Table 3.1: Methodological options for this study

(Researchers concept)

3.3 RESEARCH STRATEGY

While two types of empirical research strategies are identified as being qualitative and quantitative, the former approach uses data analysis tools which are structured in a way that attempts to maintain the highest degree of objectivity as the results are generalised and prove to be accurate (Creswell, 2013). The latter even though the researcher is responsible for stimulating certain conditions for the purpose of recording the results, the objectivity of the above approach is based on the premise that the researcher is detached from the research subjects and is therefore not able to influence the outcome of the results which are expected to remain the same if the study is repeated. This is further qualified by another scholar who emphasised the objectivity of this approach when it was stated that the same result is attainable when the researcher detaches himself or herself from the research subjects to avoid influencing the outcome and further said that the approach is used for data that could be generalised such as evaluation of outcomes and decentralisation of human behaviour (Creswell, 2013). The research strategy adopted for this current research is 'action-research' within the case study domain, derived from the interpretivists and ontological assumptions in accordance with Creswell (2013).

For the purpose of the current research, the researcher reviews the current risk management tools in use for construction projects and this is done by introducing specific stimuli to different project activities where the results are carefully observed and recorded for further analysis. The format followed in this research is done in stages as depicted in Table 3.2 below;

STAGE	ACTION RESEARCH
ENTRY	Client or researcher presents problem. Mutually agreed goals
CONTRACTING	Business and control contracting. Mutual
DIAGNOSIS	Joint diagnosis. Client data/researcher's concepts
ACTION	Feedback. Dissonance. Joint action plan. Client action with support. Published
EVALUATION	New problems emerge. Recycles. Generalisations emerge
WITHDRAWAL	Client self-supporting

Table 3.2: Action research sequence at each stage
SOURCE: (Gill and Johnson, 2010)

The stages depicted within the action research sequence in Table3.2 are explained in detail under Figure 3.4 on page 64 which discusses the 'Spiral nature of Action-research'.

3.4 RESEARCH PHILOSOPHY

There are four philosophical assumptions underlying research and these are, Ontology, Epistemology, Axiology and Methodology (Creswell (2013). In this study, 'Ontology' is the philosophical assumption has been adopted for use. Ontology is the nature of reality and its characteristics and consists of what reality is thought to be really like (Patton, 2002; Winter *et al.*, 2006). While ontology may be adopted which regards projects as a thing, another contrasting ontology may be chosen which considers a project as a process (Patton, 2002; Winter *et al.*, 2006). Modern ontology which has come to replace Naturalism is Scientific Realism which is also referred to as transcendental realism, rational realism, critical realism and empirical realism (Patton, 2002; Winter *et al.*, 2006; Moses and Knutsen, 2012). The four research philosophical assumptions are briefly in the following discussion.

3.4.1 *Epistemological Assumption*

The epistemological assumption, also referred to as reality is divided into two categories namely, Interpretivism and objectivism. The former is also referred to as naturalism and it is an approach where the social world is only understood from the perspective of the individuals involved in the activities under investigation, The latter entails an approach which seeks to predict events in the social world by looking for regularities and causal relationships between the constituent elements (Gill and Johnson, 2010).

3.4.2 *Ontological Assumption*

The ontological assumption, also referred to as 'creation' is a social science approach which studies the essence of phenomena and the nature of their existence and considers that the phenomenon of reality exists independently of the cognitively structures of observers while the epidemiological realism considers that reality to be cognitively accessible to observers (Gill and Johnson, 2010; Creswell, 2013).

3.4.3 *Human Nature Assumption*

This assumption which is also called relationship between man and society is premised on the subjectivist and objectivist approaches to social science. The subjectivists' approach relates to the voluntarism in which the individual is regarded as being completely autonomous with a free-will while the objectivists' approach is based on determination and entails the individual is regarded as being completely determined by a situation or environment in which they are (Gill and Johnson, 2010).

3.4.4   Axiology Assumption

The axiology assumption, also referred to as aesthetics, ethics and justice is also based on either the subjectivist or objectivist approaches to social science where the subjectivist approach is value biased in which research is value laden and completely subjective whereas the objectivists' approach is value neutral where research is value free and completely objective (Creswell, 2013).

3.4.5   Methodological Assumption

The approach for the methodological assumption is Ideographic, which is based on the subjectivist approach to social science in which the social world is only understood by obtaining first-hand knowledge on the subject being investigated and the nomothetic, based on the objectivist approach which emphasizes the importance of basing research on systematic protocol and technique epitomized by the approaches and methods used in the natural sciences (Gill and Johnson, 2010). The methodological assumption also uses quantitative and qualitative methods where the former uses mathematical techniques to identify facts and causal relationships while the latter focuses on determining what things exist rather than how many they are (Gill and Johnson, 2010; Creswell, 2013). The research philosophy adopted for this study which was undertaken within the action research domain, where the researcher applied the qualitative approach within the methodological assumption to implement project management strategies for risk management.

The five assumptions in terms of this study involved the introduction of stimulating actions to project activities with a pre-determined intention to record the outcomes. This is done in order to evaluate the process conducted within the stages of the value management cycle. It is also in line with the interpretivist and positivist approaches derived from the Ontological and Epistemology assumptions which entail the application of different methodologies to tackle the research. The epistemology approach adopted for this study was inherently based on the assumptions that project initiators are not aware of value management, potential risks on projects are mainly due to the lack of involvement by project managers from the initiation stage, risk management is not adequately provisioned for during the planning stage, project managers on most projects lack formal qualifications in project management, contingencies and float are the most used risk mitigation factors on construction projects and most projects commence without a project charter. The current research aligns itself to the interpretivist research paradigm because of its qualitative, subjective and humanistic nature. Further, this paradigm is aligned to the interpretivist method because the nature of this current study is that of action research (Patton, 2002; Winter *et al.*, 2006; Moses and Knutsen, 2012).

3.5 RESEARCH APPROACH

The research approach enables the researcher to distinguish between different social science research methods which are deductive and inductive by nature (Gill and Johnson, 2010). There are predominantly three approaches to research and these are 'deduction', 'induction' and a combination of both approaches as presented in the following discussion.

3.5.1 Deductive approach

This is a method which entails the development of a conceptual and theoretical structure prior to its testing through empirical observations and the process of deduction is divided into three different stages namely; concepts, rules and operationalisation (Gill and Johnson, 2010). For the purpose of this research, the stages presented are complied with in that the researcher decides which concepts represent important aspects under investigation, followed by the setting of rules for making the observation and determining when an instance of the concept has empirically occurred and finally creating the measures which represent the occurrences of the concepts under investigation. This is done in terms of the construction project risk-related problems experienced which lead to project activities not being completed within cost, time and quality and are further assumed to be caused by the non-compliance with some project management techniques within the value management domain during the planning phase. It was from this logic that the researcher re-engineered some risk inherent activities by introducing the project management strategies and made observations which were recorded and documented for analysis. The deductive approach enables a researcher to organise grounds into patterns that provide conclusive evidence for the validity of a conclusion (Palys, 2003).

3.5.2 Inductive approach

Although this approach is different from the deductive approach, it is relevant to this research and it is important to briefly discuss it in order to provide the reader with an idea of what the approach entails. An inductive approach focuses on developing premises that are true and one must begin with a true premise in order to arrive at true conclusions (Palys, 2003). In this study, a true premise which the researcher has developed is that the use of project management strategies during the entire project life cycle results in the elimination of most project risks and is employed as an effective risk management tool. The task of a researcher is to interpret and analyse gathered data with the aim of formulating a theory because it is from this approach that theory follows data and not vice versa as in the deductive approach (Palys, 2003). Similarly, the processes undertaken during this research created a set of theory upon which data were gathered.

The induction approach is also explained as the reverse of the deduction approach which involves moving from the 'plane' of observation of the empirical world to the construction of explanations and theories about what has been observed (Gill and Johnson, 2010). This is in sharp contrast to the deductive tradition where a conceptual and theoretical structure is developed prior to empirical research where theory is the outcome of the induction approach. The current research employed a combination of both the deductive and inductive approaches to research. This is because this study commences with the analysis of literature, identifies a single problem and isolates the major research question while listing assumptions drawn from such literature. Data were then collected which inform the explanations and theories from the observations.

3.5.3 Paradigmatic approach

Paradigms are systems of interrelated ontological, epistemological and methodological assumptions (Terre Blanche *et al.*, 2006). They act as perspectives that provide a rationale for a study and commit the researcher to particular methods of data generation and interpretation (Wellington and Szczerbinski, 2007). The paradigmatic beliefs of the researcher influence the purpose of the research, how it is conducted, how the role of values and ethics in their work will be assessed, how the formation of relationships with research participants will be done and how their work will be presented (Wellington and Szczerbinski, 2007). The above requires that the researcher ensures coherence in the study design because the design has an impact on the nature of the research question and the manner in which the research will be conducted. Coherence is maintained by ensuring that the research question and the research methods adopted fit logically within the paradigm used (Terre Blanche *et al.*, 2006). For the purpose of the current study, the researcher opts for the paradigmatic standpoint of critical theory as a metatheoretical paradigm and selects the action research (AR) design as a methodological paradigm.

3.5.4 *Metatheoretical paradigm: Critical theory*

Critical theory originates from the writings of Karl Marx although it is usually associated with the Frankfurt School as Marx's principles are said to be concerned with social equality, equity and justice (Shook and Margolis, 2006). Scholars assert that Marx's socialist theory was an attempt to reconstruct society in a more just way (Shook and Margolis, 2006; Kettley, 2010). Critical theory was the preferred choice for this study because it entails working with various project stakeholders in the areas where risk is eminent on the project in order to achieve project success in terms of delivery within the constraints of time, cost and quality (Shook and Margolis, 2006).

The term 'epistemology' refers to the nature of knowledge, while the term 'ontology' refers to how individuals perceive and act on this knowledge (Zuber-Skerritt, 2011). The epistemology of critical theory suggests that people's perceptions and expectations of reality are the object of a study (Guba and Lincoln, 1994). Therefore, epistemology implies that the researcher needs to interact with people, engage them in the investigation, observe events and discover the lived reality of people and that the ontology of critical theory is that knowledge is socially constructed (Guba and Lincoln, 1994). The epistemology and ontology of critical theories create a perfect link to action research, the methodological paradigm adopted for this study The reason for this is that they allow participants to be engaged in the research process, to plan and to take action to improve their circumstances through a systematic process of critical reflection on the current situation (Gill and Johnson, 2010).

3.5.5 *Critical Social Research*

The Critical Social Theory (CST) is a school of thought within the domain of Critical Social Research (CSR) with its primary objective being the improvement of human conditions and its focus on general theoretical problems, as well as specific investigations of concrete problems of contemporary social organisations (Myers and Avison, 2002). Critical Social

Research as shown in the three research paradigms presented in Figure 3.1 is underpinned by a critical dialectical perspective that attempts to dig beneath the surface of historically specific, oppressive and social structures ((Goede *et al.*, 2012; Harvey, 1990).

Critical social theorists view knowledge as being structured by existing sets of social relations (class, gender or race) that are oppressive. The fact that knowledge is critique, it is a dynamic process not a static entity and it is also a process of moving towards the understanding of the world and of the knowledge that structures perceptions of the world (Harvey, 1990). The aim of critical social research is to emancipate or change suppressive behaviour.

Critical social theory is meant to break with traditional hypothetical deductive methods, which are oriented towards the preservation and gradual reformation of the status quo (Myers and Avison, 2002). Although Critical social theory is intended to be a radically different approach that takes into account the human construction of social forms of life and the possibility of their recreation, others argue that the primary difference between traditional social theory and critical social theory is the researcher's attitude towards his/her world and work (Myers and Avison, 2002).

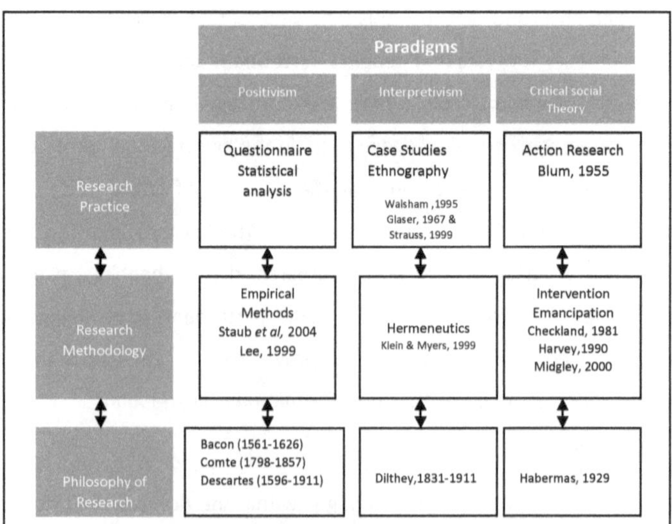

Figure 3.1: Summary of the three research paradigms
SOURCE: (Goede *et al.*, 2012)

3.6   RESEARCH MIXED METHODS

The methodology used for this study is a combination of both the qualitative and quantitative approaches which involves techniques such as observation and case study. This is also referred to as 'mixed methods'. The mixed methods approach has its origins in psychology studies in which researchers used their multi-method matrix to look at various data collection techniques in which qualitative methods such as observations and interviews were used together with quantitative methods such as an embedded survey within a case study design and traditional surveys (Sieber, 1973). Although the approach was not used in this study, the choice of the embedded survey within a case study design is illustrated in Figure 3.2 below.

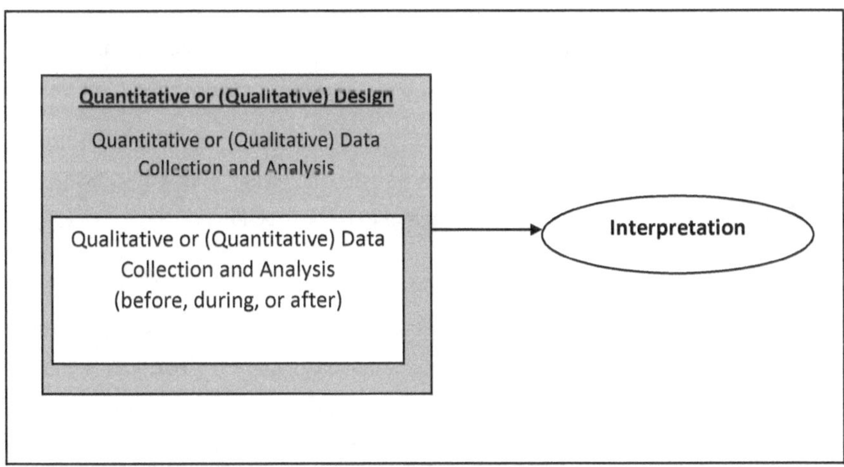

Figure 3.2: The Embedded Design
(Sieber, 1973; Creswell and Poth, 2017)

Some researchers state that both qualitative and quantitative data collection methods are used and results are compared to see if the outcomes are the same which in turn provides a means of embedded design (Jick, 1979). The mixed methods choice is preferred because the type of research undertaken is one which looks more at people's behavioural tendencies with regard to the decisions and choices of risk management tools used on a project. The strategy to opt for both the qualitative and quantitative approaches is further informed by supporters of both who claim that the strategy chosen must be described procedurally and generically irrespective of whether the approach used is 'qualitative' or 'quantitative'

(Punch, 2000). The methodologies used in this study are briefly described in the following discussion.

3.6.1   Qualitative Research

Qualitative research is more intimate and seeks depth rather than breadth (Ambert *et al.*, 1995). Qualitative research is further described the type of research which employs the use of quantitative data, for example interviews, documents, observations and stories, usually to understand and explain social occurrences (Goede *et al.*, 2012). The emphasis is on processes and meaning that cannot always be measured in terms of quantity, amount or frequency. It includes words, images, and sound

The current research was also based on that premise as a more intimate type of study which seeks depth rather than breadth. This approach is further used because it has been extensively used in the examination of the effectiveness of public sector delivery (Ambert *et al.*, 1995). It is for these reasons that this research is classified as a public sector study because it attempts to determine the potential benefits of using project management strategies as a contributor to effective risk management on construction projects.

3.6.2   Quantitative Research

Qualitative research is used to study natural phenomena which includes experiments and numerical methods. It is data based on numbers or values (Goede *et al.*, 2012). This is the main type of data generated by experiments and are primarily used by positivists, but can also be used by interpretivists or critical researchers. Additionally, the reason for the choice to use both approaches in this study was to enable the different aspects of this current research book to be covered. Scholars assert that the qualitative and quantitative methods are used because they both subscribed more to a phenomenological, holistic, non-subjective, process oriented and social anthropological world view (Cook and Reichardt, 1979).

For the purpose of this study, the methodology employed is adapted from the action research cycle in which six stimuli are introduced to project activities through the phases of the action research cycle facilitated by the researcher. The identity of the project was withheld to ensure anonymity and only referenced as 'The Project'. The six actions below were introduced as stimuli on project activities and the results recorded for analysis. They are the basis upon which the action research for this study is premised and also presented in detail as Annexure A;

Action 1:   *Signing of a Memorandum of Understanding (MoU)*: An agreement in form of a MoU was signed with the professional team. A sample of the MoU is presented as Annexure J.

Action 2:   *Implementation of a procurement plan*: A procurement plan was put in place for the procurement of sub-contracted services and materials as well as Plan and critically supervise project logistics by the researcher in his capacity as project manager.

Action 3:   *Audit of skills and competencies*; The project manager audited the contractor and sub-contractors labour for skills and experience.

Action 4:   *Implementation of a Stakeholder Management plan*: Identification of stakeholders, their needs and project expectation.

Action 5:   *Implementation of the team induction and orientation plan*: Development of a human resource strategy retention on the professional team.

Action 6:   *Contract management*: Review of the qualifications and skills of the contractor and the critical evaluation of compliance documents submitted by the contractor.

3.6.3. Reasons for each action taken

A comparison of quasi-experiments, true experiments and action research revealed that quasi-experiments involve the analysis of events that have naturally occurred without the intervention of the researcher. True experiments and action research both entail the analysis of the researchers' direct intervention (Gill and Johnson, 2010). The action research intervention was based on the criterea chosen for this study where different stimulating actions were introduced to project activities as a direct intervention of the researcher. The stimulating actions introduced to the project are explained in detail in the following discussion.

3.6.4 Action 1: Signing of a Memorandum of Understanding (MoU)

A Memorandum of Understanding was signed with every consulting firm in this study which comprised the professional team. This was done in order to ensure that the same professionals assigned to the project by the consulting firms which were awarded the contract in a particular field to be retained for the duration of the project. This is because research stated that a high turnover of project staff is one of the causes of delays and delivery of sub-standard quality on projects. This occurs when duly qualified professionals initially assigned to the project are replaced by less qualified persons who do not understand the project, subsequently this affects the project time and quality related some risks (Enshassi and Mosa, 2015). Therefore, the process of signing the MoU adapted from the phases within the action research cycle is able to better manage the risks.

3.6.5 Action 2: Implementation of a procurement plan or strategy

The literature referenced for this study revealed that some projects experience delays while others are prematurely terminated or abandoned because of various reasons which include the late ordering of materials, wrong materials ordered and the turn-around time for some of the materials not taken into account based on the activity for which such material is needed (May and Govender, 1998). In addition, some material ordered was found not to be in accordance with the specified standard and therefore accounted to an inferior quality of the end-product (May and Govender, 1998). Inexperienced and less qualified contractor and sub-contractors' manpower is also identified as a contributing factor to delays and quality related risks (Atkinson, 2007). The implementation of an implementation of a procurement plan were undertaken from processes adapted from the phases within the action research cycle and implemented with a view to manage project schedule and quality related risks.

3.6.6 Action 3: Implementation of the skills and competence audit on the project

Some researchers also noted that existing contractors lack the necessary skills to manage finances and correctly price the Bill of Quantities in tender documents which leads to under-pricing and ultimately causes numerous project risks (Baloyi and Bekker, 2011). In certain instances, projects awarded to BBBEE beneficiaries are either delayed, abandoned, completed and delivered below the set standard or terminated prematurely due to lack of expertise in the contractors. This prevents such contractors or Sub-contractors from duly complying with bid requirements and specifications (Jansen and Christie, 1999). The reason for implementing action number 3 is to ensure that an audit of the contractors' skills and competencies is compared and verified against the reported level of skills, competence and experience as prescribed in the bid document which formed the basis for the award of the contract. The implementation of the skills and competence audit was undertaken using processes adapted from the phases within the action research cycle, which also aimed at ensuring that risks related to competence and skills on the project were identified and managed before any of the project objectives or deliverables was affected.

3.6.7 Action 4: Stakeholder management

The Project Management Body of Knowledge (PMBoK) places an emphasis on stakeholder management in which influential stakeholders and stakeholder expectations must be identified and appropriately managed (PMI, 2015). Contract scope changes, lack of understanding the needs, expectations and abilities of the different project stakeholders is a cause for numerous project risks. These include unplanned scope changes, which have the potential of affecting the schedule and budget and also unrealistic project expectations by some of the stakeholders, understanding the project needs of some sectors of stakeholders so that the project is relevant, understanding the ability of the sponsor and client regarding influence, authority and financial limitations. During this phase of stakeholder management, the community within the project vicinity was also engaged and their needs and expectations managed. This was essential because some projects were shut down by protesting community members who felt excluded from project employment opportunities as many people within the surrounding communities were unemployed and demanded to be employed within the project (May and Govender, 1998; Atkinson, 2007). The stakeholder management action derived from the phases within the action research cycle was implemented for the purpose of managing all stakeholder related risks.

3.6.8 Action 5: Team induction and orientation plan

Team induction was intended to introduce old and new project members and provide them with an orientation and overview of the project. It was also meant to appraise and update all project members on the next activity in sequence to be undertaken and the various tasks which the activities entail. In addition, induction was done so that in an unlikely even that there was a new member on the project team or a new sub-contractor, it would assist to align the new comers to the project deliverables in order to attain 'unity of purpose' where the different activities ultimately translate into the same project objectives and deliverables (Enshassi and Mosa, 2015). It was also meant to bring all project stakeholders abreast with everything that was happening on the project as well as the envisaged next set of activities so that every project team player is able to plan in the short-term for the respective individuals' role in the overall project value chain. Induction eliminated risks associated with different role players implementing their own individual work plans which do not conform to the overall programme set by the project manager.

3.6.9   *Action 6: Contract Management*

The contract management action introduced, was mainly aimed at the tender document with specific reference to the Bill of Quantities (BoQ). The intention was to ensure that the tendered prices are balanced and market related. In accordance with the literature review, some contractors classified as Previously Disadvantaged Individuals (PDI) were reported to lack tender pricing knowledge which resulted in the allocation of higher rates to lowly rated items and lower rates to higher rated items. It was suggested that this was one of the risks associated with improperly priced documents because some of the affected contractors subsidise the project because commodities supplied to the project were quoted at a lower rate while the market rates for the same commodities were higher than the quoted price (Ngwenya, 2007; Baloyi and Bekker, 2011; Enshassi and Mosa, 2015). Price related risks have the potential of rendering some contractors bankrupt and a cause for abandoning the project site.   This was the reason why this study deemed it important to implement the contract management action.

3.7   ACTION RESEARCH AS A PARADIGM AND METHODOLOGY

A research design is a formal plan used to conduct a study and it specifies exactly how the study will be conducted (Mertler, 2009). It is critical to carefully conceptualise the design of a study prior to its commencement, considering all aspects of the study and how they fit with the paradigm of the researcher, as well as the nature of the research question. The research design aspects include not only how data will be generated and analysed but also primarily how the study will be planned (Mertler, 2009). The research design selected for the current study is presented in the following discussion.

The researcher selected action research (AR) as a methodological paradigm for this study because besides being a methodology, it is also a paradigm, based on values that promote the social good (Zuber-Skerritt, 2011). In the course of an investigation, action research allows participants to experience insightful, emotional moments and personal growth akin to the transformational moments that occur in action learning (O'Neil *et al.*, 2003). Action research emphasises cognitive and operational changes in participants, which results in a better understanding of social reality, rather than the researcher making assumptions about how things are and how action and attitudes may be adjusted (O'Neil *et al.*, 2003). It is by its nature, directive and interventionist, trusting that people as social beings are generally motivated to self-organise and to work collaboratively, contributing all they can in unpretentious ways to the common good (O'Neil *et al.*, 2003).

Another reason why the researcher chose action research is because it links with critical theory as it also places a high value on the democracy of the research process, allowing for free and full participation of the participants (Stringer, 2008). It became relevant for the researcher to use action research for this study because it enabled the researcher to work collaboratively with the participants, regarding them as co-researchers, raising their awareness of the need for change and helping them to visualise and actualise change. The epistemology and the ontology of critical theory resonate with the participative and democratic assumptions of action research (Kindon *et al.*, 2010). Some scholars view action research as a research methodology which combines research, education and action (Collins, 1999; Kindon *et al.*, 2010). Action research is a form of participative, person-centred inquiry which allows the researcher to conduct research with people, as opposed to research on them or about them and also helps them to change and grow through the process (Wellington and Szczerbinski, 2007). It is viewed as an emancipatory approach to knowledge production and utilisation, aiming to actively involve oppressed people in the collective investigation of reality, in order to transform their knowledge (Collins, 1999; Creswell, 2008; Kindon *et al.*, 2010).

As indicated in Chapter 1, the intention of the researcher was to facilitate an improvement in the successful delivery of projects within the project success criteria of time, cost and quality and implement project management strategies to improve risk management which accounts for many challenges on construction projects. In order to achieve this, the researcher adopted dialogical, dialectical and hermeneutic approaches in the interaction with all relevant stakeholders who were participants as propagated by Collins (1999), Creswell (2008) and Kindon *et al.* (2010). The dialogical and hermeneutic approaches in research refer to a more democratic empowerment approach which seeks full participation and it also engages participants as equals (Guba and Lincoln, 1994; Stringer, 2008). These approaches require them to think critically as action research provides a simple but powerful framework which engages people in critical thinking (Kindon *et al.*, 2010). Figure 3.3 below indicates the process followed in this study.

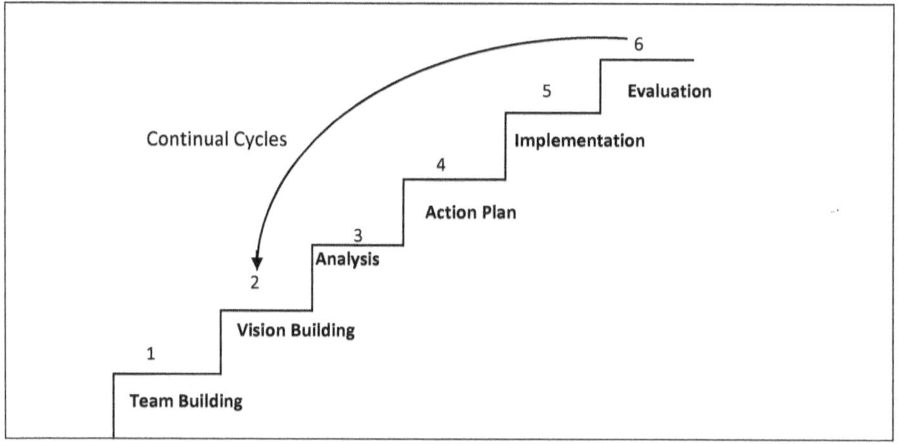

Figure 3.3: The Action-research process

(Zuber-Skerritt, 2011)

The stages within the action research process depicted in Figure 3.3, which is the action-research process were adopted in this study to facilitate the interaction between the researcher (Project manager) who in the above scenario is the facilitator and the participants. While stages 1, 2 and 6 in Figure 3.3 are not covered by the usual action research cycles as depicted by some scholars in their spiral of cycles, these stages are important in the action research process (Guba and Lincoln, 1994; Stringer, 2008). In terms of the current study, stage 1 involves creating a good relationship between the researcher and the participants and among the participants themselves. Stage 2 involves the development of a common vision by the team and this will carry them through to stage 6 which involves the evaluation of the achievements of the research. The action research process constitutes a spiral of cycles, each cycle consisting of four phases (Guba and Lincoln, 1994; Stringer, 2008) indicated by stages 3, 4, 5 and 6. Stage 3 which is planning and context analysis, involves the identification of the problem; stage 4 involves the development of a strategic plan to improve the situation; stage 5 involves the implementation of the plan and stage 6 involves the evaluation of the action plan as well as critical reflection of its impact and significance for further change.

At the completion of each cycle (stages 3-6), participants reconsider the situation for a review, reflect for reanalysis, and re-act to transform their actions (Wellington and Szczerbinski, 2007; Stringer, 2008; Turesky and Gallagher, 2011). The participants continue to reflect on, and re-act their situation, proceeding from one cycle to the next, until all stakeholders led by the project manager / researcher are satisfied that the desired change has been achieved in accordance with Figure 3.3 (Kindon *et al.*, 2010). The spiral nature of the process undertaken in this study is depicted in Figure 3.4. The illustration presented shows that the action taken in this study will take the form of a spiral cycle where stimulating action will be introduced and the outcome observed. Where the outcome is not in accordance with the intended result, the process will be repeated until the desired outcomes are realised. During the first process within spiral 1, the researcher plans the intervention, then, acts on it, observes the outcome and reflects on the process. When the outcome is not in accordance with the intended result, the process is repeated until the desired result is achieved at process number 3.

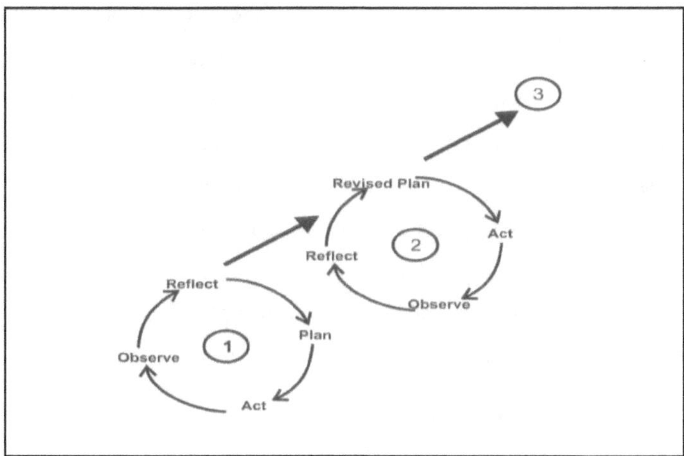

Figure 3.4: The spiral nature of Action Research
(Zuber-Skerritt, 2011)

The entire research process necessitates that participants reflect on and experientially learn from their actions and draw from their own experiences (Turesky and Gallagher, 2011). In order for this to happen, participants are engaged in reflective and dialogical conversations (McIntosh, 2013). The project manager, who is also the researcher in this study engaged in collaborative dialogues, through which all stakeholders gained access to overall vision of the project through which wider opportunities for understanding could unfold (McIntosh, 2013). The project manager requires to collaboratively work on issues that are potentially risky to the project by using the phases of the action research cycles. This will require the project manager to define issues, ideas and government policies about the roles of different stakeholders on the project, so that they are all able to develop a deeper understanding of how their respective roles fit into the overall project success picture (Collins, 1999; Creswell, 2008; Kindon *et al.*, 2010). The Project manager collaboratively identified problems which accounted for risks on the project which had the potential to hinder the ability to achieve project success within time, cost and quality. This framework, where participants were required to look or observe, think and act in working out their situation demanded full commitment by the participants from the beginning of the project to the end (Kindon *et al.*, 2010).

In traditional research, learning is assumed as an outcome which in reality is rarely reflected upon by participants and opportunities to learn are restricted to the findings of the data analysis (Mertler, 2009; Kindon *et al.*, 2010). In action research, participants learn throughout the research process (Kindon *et al.*, 2010). This framework promotes the values of democracy, equality and liberty, which are the social values of action research. This value-based design implies that participants' views are respected, they are equals and they participated freely and fully in the investigation, looking into the situation that needed to be changed, thinking of solutions, and acting together to implement solutions to bring about change. Compliance of the participants to the set standards by the project manager or researcher was a requisite in order for the desired outcomes to be achieved.

3.8   SYNTHESIS OF THEORY INTO THE VM METHODOLOGY

The intention of this research was to employ the strategy of value management as an effective contributor to project risk management for construction projects. In view of the views and techniques expressed by the different scholars cited in the preceding paragraphs, this study proposes value management as a methodology for the identification and management of risk on construction projects which is employed on the project being investigated. The methodology involves the use of value management:

(i)   By involving the Project Manager from the project conceptualisation phase until the end of the project in accordance with the integrated approach to VM (Goede, 2018).

(ii) By getting the project manager to develop the project charter with realistic objectives based on the requirements of the client, sponsor and user with input from all project stakeholders (PMI, 2015).

(iii) To conduct an analysis of the project delivery process including traditional competitive bid (Jansen and Christie, 1999).

(Iv) By using the integrated approach to soft value management phases of planning, learning, analysis, creativity, evaluation and development to identify and manage risk on projects (Al-Yami, 2006).

(v)   To identify risk by employing steps such as; firms' capability and experience, specific client attributes, construction industry factors, Constraints on quality, time and cost as well as forces external to design and construction derived from the Risk identification hierarchy for construction projects (Zhi, 1995).

(vi) To identify project risks which might undermine project quality and plan for such risks before they manifest to affect deliverables and objectives according to the risk identification hierarchy process (Zhi, 1995; PMI, 2015). These may include a critical analysis of contractors and sub-contractors so that the required level of expertise is retained throughout the project.

(vii) To obtain acceptance and active support of both the engineers, contractor and the professional team in (PMI, 2015),

(viii) To provide a link between the agreed design and the final product and prevent cost over-runs which are a result of re-work in terms of stakeholder management (PMI, 2015)..

(ix)   To establish and manage the value system of the client and ensure that the clients' values are communicated and understood by the project team in terms of stakeholder management requirements (PMI, 2015).

(x)   To optimize the function of project assets at lowest cost in accordance with the assertions of (Mandelbaum, 2006)..

(xi)   To identify and manage stakeholder and community related risks (PMI, 2015).

(xii) As an organised approach to get more from investment and eliminated anything that added cost but did not affect the basic performance (Mandelbaum, 2006).

(xiii) To look at various methods and their associated costs and selecting the cost-effective method.

3.9   RESEARCH DESIGN

The current research was a case study conducted on a school construction project in the Ngaka Modiri Molema District Municipality (NMMDM) of the North West Province. It is based on action research undertaken by the researcher who is employed as Project manager on the project. The study is designed to review risk management tools and techniques and the implementation of project management strategies within the value management domain for risk management on the project. As such, the study further seeks to fully understand how the strategy is to be implemented in order to obtain the best results.

This study will also assess the way activity tasks are planned and executed within the context of the project and also examine the procedure of how the planning and organising of tasks is done during this phase in order to ascertain whether activities which are not properly planned account for greater risks on the project. Stakeholders on the project include the professional team (consultants from various professional disciplines), contractor, sub-contractors, project sponsor and user client as the sample. Scholars assert that research design is described as a general plan of how the research objectives that had been set would be achieved and the three main strategies for research design are said to be, Case study, Survey and Experiment (Sauders *et al.*, 2003). For the purpose of this research, only the case study was discussed as the experiment and survey research designs were not used and therefore irrelevant for the current purpose.

3.9.1 Experimental research design

The experimental research design adopted in this study critically examined some of the problems faced by those engaged in deductive research such as matching experimental and control groups and the biases which might arise from experimental research design (Sauders *et al.*, 2003). The researcher critically examined the problems currently experienced in relation to the project in which most of the activities were behind schedule, which caused a delay in the entire project. The researcher used control groups matched with the specific experimental roles of the focus groups to examine the outcomes. This process of matching experimental groups prior to any treatment was important because it allowed for some confidence regarding the internal validity of any consequent findings (Sauders *et al.*, 2003).

3.9.2 Research Site and participants sampling

Prior to the implementation of one of the project management strategies used in this study in the form of the integrated approach for value management through carefully selected stimulating action introduced to project management processes by means of action-research, the overall progress of work on the project has been low and most activities and tasks are behind schedule. The project, which entails the construction of a school was scheduled to be completed within 16 months from the commencement of 1st April 2016 to 30 August 2017. At the time of the intervention by the researcher, in April 2018, the project was in default and had been placed on penalties pending termination. Documents to support the report in terms of contract duration and default letter are presented as Annexures C, and D, being the letter of default and document indicating the initial contract duration respectively.

During the action research component of this study, the researcher, in his capacity as project manager applied for extension of time from the employer through the client in a letter presented as Annexure E. The application was approved by way of a letter presented as Annexure G in which an additional 150 days were granted. It was during the extension period that the project management strategy was implemented and after an effective 90 days, a progress assessment on the project work revealed an increase from 15% to 80% in September 2018. this was achieved through the implementation of action recorded on the sample document presented as Annexure A, being stimulating action introduced to management processes, Annexure B being the 'Research process diagram', Annexure I, being the 'Data collection sheet'. The practical completion was set for 1st February 2019 while the completion of works was scheduled for 29th March 2019, before the project handover on 1st April 2019.

The focus of the study is on a school construction project within the Ngaka Modiri Molema District of the North West Province at which the researcher was Project Manager as it was a Case Study. The North West is one of the nine provinces in South Africa experiencing many protest actions by citizens because of service delivery and the majority of these are due to poor or lack of school infrastructure of acceptable standards in communities (Akintoye and MacLeod, 1997; Booysen, 2007).

In view of limited resources and time constraints, it was impossible to have a large number of participants who would potentially inform the research process. Therefore, the researcher purposively selected a smaller group of role-players and members of the professional team whose roles on the construction site constitute significant input to project deliverables to provide information on which the study was grounded as per Collins (1999), Creswell (2008) and Kindon *et al.* (2010). The researchers initial aim was to work with all people on the project but due to the fact that the roles of some were not relevant to the study and that their inclusion would amount to the generation of unnecessary data which would be unusable, he opted to include only those participants whose roles were relevant to the project in terms of the requirements of the study. Due to the small number of participants currently active on the project, the researcher extended the study population to the role-players who had once been part of the project but left due to different reasons. The aim was to widen the scope of participants and also to get the views of those who had left the project.

The fact that the project site was a school situated in a smaller socio-economic context and faced with similar challenges to schools of a similar nature, the researcher provided the participants with a brief description of the study and what was required of them. The research site for this study was a building project for the construction of a high school where ten blocks of classrooms, an administration block, a science laboratory, a computer laboratory, a multi-purpose kitchen and dining hall, a library, separate ablution blocks for boys and girls, sports facilities and a bulk biosorp septic tank were to be erected. It can therefore be stated that the researcher used purposive sampling to select participants for this study, sampling technique which seeks to ensure that the diverse perspectives of people who are likely to affect the problem under investigation are included in the study (Patton, 2002; Neuman, 2011). Purposive sampling is meant to select participants for a variety of purposes, namely maximum variation sampling (which seeks to include people who represent diverse perspectives in any social context), extreme case sampling (which strives to include particularly troublesome or enlightening cases), typical sampling (which endeavours to include participants who are typical of people in the setting), or theory or concept sampling (which tries to include participants who have particular knowledge related to the issue studied) (Collins, 1999; Creswell, 2008; Kindon *et al.*, 2010). The researcher further used typical sampling to select the current site for the study because of the belief that both the participants and the setting were most typical of its population and also resonates with the research topic under investigation (Turesky and Gallagher, 2011).

The choice of this project was also due to the background knowledge which the researcher had about the project dynamics faced on the site as project manager and also because the persistent backlog of successfully completed projects in the district and South Africa as a whole was an increasing cause for concern (Turesky and Gallagher, 2011). The intention to work with stakeholders on the project was also aimed at helping them to develop ways of improving their project management abilities when executing their respective tasks so that the successful completion of tasks within time cost and quality was appreciated. However, the researcher had to involve other role-players who were once part of the project in the study. Due to the emerging and dynamic nature of action research, the researcher decided to conduct research at the site where the participants' free will to be part of the study was obtained as ethics require that no one is forced to take part in a study if such persons are not willing to participate (Turesky and Gallagher, 2011). As a result, the researcher conducted the study using focus group workshops with stakeholders who were available (see Chapter 4). In total, fifteen participants were divided into three groups, referred to as focus groups and workshops are conducted for each group facilitated by the researcher. Focus group 1 was comprised of engineers, the Quantity Surveyor (QS) and contractor drawn from the project role-players, Focus group 2 was made up of engineers, sub-contractors and architects, while focus group 3 was comprised of sub-contractors, social facilitators and community liaison officers. Table 3.3 provides biographical information of the participants.

FOCUS GROUP 1				
Participants	Age category (Years)	Gender	Race	Experience (Years)
Architect 1	50 - 55	Male	Black	20 -25
Engineer (Structural)	40 – 45	Male	White	10 – 15
Engineer (Civil)	40 -45	Female	White	15 – 20
Quantity Surveyor	35 – 40	Female	Black	5 - 10
Contractor (Foreman)	60 - 65	Male	Black	35 - 40
FOCUS GROUP 2				
Participants	Age category (Years)	Gender	Race	Experience (Years)
Engineer (Electrical)	45 – 50	Male	Black	10 -15
Sub-contractor (Civil)	55 – 60	Male	Black	0 – 5
Sub-contractor 2 (Electrical)	30 - 35	Male	Black	15 – 20
Contractor (Site Agent)	35 – 40	Male	Black	10 - 15
Principal Agent (Architect 1)	50 -55	Male	Black	15 - 20
FOCUS GROUP 3				
Participants	Age category (Years)	Gender	Race	Experience (Years)
Community Liaison Officer	35 - 40	Female	Black	0 - 5
Social Facilitator	40 – 45	Male	Black	5 – 110
Sub-contractor (Tiles)	40 -45	Female	Black	15 – 20
Sub-contractor (Brickwork)	35 – 40	Female	Black	5 - 10
Sub-contractor (Roof)	45 -50	Coloured	Black	10 - 15

TABLE 3.3: Biographical data of focus group participants

(Researcher's concept)

3.9.3 Methods of data generation

The generation and analysis of data in action research is significantly effective when it is done as an interactive process between stakeholders (Turesky and Gallagher, 2011). Most of the data for this study was therefore generated from the interaction between the participants and the researcher, as well as from their interaction with one another. Data generation refers to the various ways in which data is obtained for a study (Turesky and Gallagher, 2011). In qualitative research, there is a wide range of methods of data generation, namely personal experience, introspection, life stories, interviews, observations, reflections, and interactions (Turesky and Gallagher, 2011).

3.9.3.1 Narratives: Cycle 1

In this study the researcher generated data using the method of narratives which are spoken or written accounts of an event or an action in chronological order (Turesky and Gallagher, 2011). The purpose of narratives is to convey experiences as they are expressed in the lived and told stories of individuals (Turesky and Gallagher, 2011). The researcher requested the participants in focus groups to write narratives (participants' narratives) with the purpose of investigating their individual perspectives on project task management and to explore the individual issues that they face that hinder their ability to complete their activities or tasks in time. The participants were required to write a narrative about what is good and what is challenging in the whole project value chain, recorded in a memo presented as Annexure F. Allowing the participants to write narratives about their experiences helped the researcher to obtain insight into their understanding of sequencing project tasks and activities and to establish a starting point for working with them. Apart from narratives, the participants did drawing narratives. This exercise was meant to stimulate the participants' thinking by allowing them to visualise the ideal causal linkages they wanted to work with and to think of how they could plan to achieve their desire

3.9.3.2 Recording of action learning sets: Cycles 1 and 2

An action learning set is a small group of three to five members working together in a specified area of discussion facilitated by a researcher(Turesky and Gallagher, 2011). The participants were engaged in action learning sets that were underpinned by the action learning principles of capacity building and continual improvement of professional practice.

It also included mutual respect and commitment to establishing and pursuing goals as well as working together to achieve them (Turesky and Gallagher, 2011). In this kind of arrangement, the participants were assisted to improve their performance and achieve their goals (Buys, 2010). The researcher in his role as facilitator, acted as a coach to facilitate the improvement of their instructional leadership (Turesky and Gallagher, 2011). In this regard, coaching was focused on helping team members to improve their problems as well as to develop and support them in their overall development for their current positions and for future functioning on the project (Buys, 2010). This was a non-directive form of coaching, which was focused on helping team members to close the gap between where they needed to be and where they were (Buys, 2010).

The participants were required to discuss the current state of their tasks or activities in relation to the overall activities of the project, set their own achievable goals that would effect change and improvement, explore new possibilities that would lead to their desired change, support one another in their plan of action and assess their own performance (Buys, 2010). The members of an action learning set help one another to improve their professional practice and leadership and this constituted an ongoing and sustainable professional development activity. Figure 3.5 illustrates the process undertaken by the researcher in his capacity as a coach and also presents the roles of participants as learners in their action learning sets. The participants were required to discuss their current situation, set their own achievable goals that would effect change and improvement, explore new possibilities that would lead to their desired change, support one another in their plan of action, and assess their performance (Buys, 2010). The members of focus group helped one another to improve their professional practice and leadership, and this thus constitutes an ongoing and sustainable professional development activity. Figure 3.5 illustrates the process undertaken by the researcher as facilitator and also the participants as role-players responsible for decisions in their respective roles on the project.

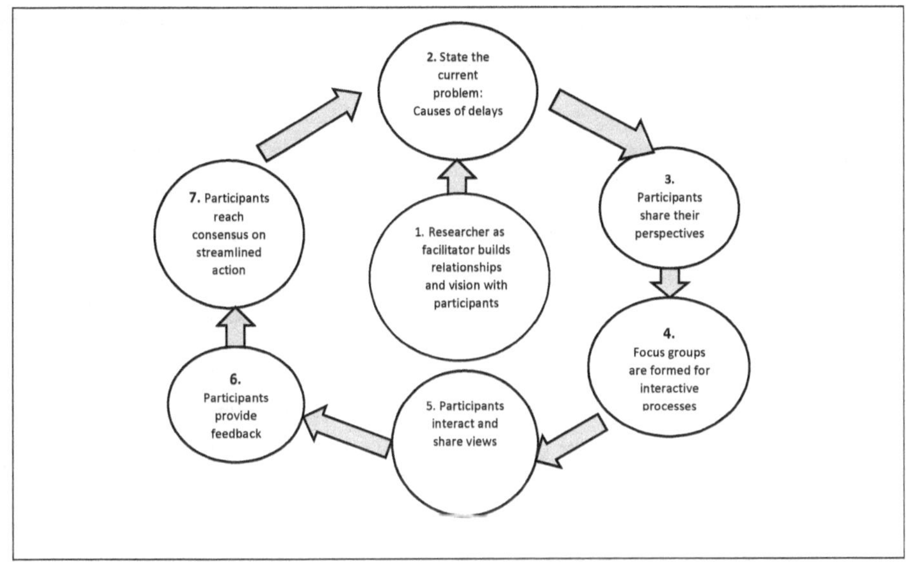

Figure 3.5: The focus group interaction process within action-learning

(Researcher's concept)

To ensure continuous and long-lasting legacy in their project roles and responsibilities, the participants met regularly as a group to critically reflect on the research process. Their meeting involved identifying and discussing issues that hinder their abilities to complete tasks on time, developing strategies to help improve the identified hindrances, discussing the implementation of such strategies, reflecting upon the effectiveness of the strategies and the significance of their changed practice for their support to the overall project goals (Turesky and Gallagher, 2011). In order for this to happen, the researcher established a good relationship with the participants and ensured that the participants, too, had a good relationship with one another. Establishing a good working relationship required that the participants openly state what they expected from their participation in this study, how they visualised working together, when and how often they interacted with the researcher, and what was expected of the participants (Buys, 2010). The researcher also ensured that the encounters were a safe and supportive environment for the workshop and further helped the participants to understand that being older and more experienced does not guarantee expertise in any particular area. He further emphasised the importance of unity of purpose to empower one another to be in the execution of their respective tasks to help them in the achievement of overall project success (Turesky and Gallagher, 2011).

In their action learning focus groups, the participants discussed issues together in a dialogical and productive relationship that created a sense of community through the sharing of perspectives, the negotiation of meaning and the development of collaboratively produced improvement strategies that they could use to improve the achievement of project deliverables (Collins, 1999). To ensure the effectiveness of the action learning sets, general principles of conducting a group discussion were followed. Participants were encouraged to respect one another, to accommodate other people's opinions, to take turns in speaking, to listen to one another, not to dominate the discussion and to treat others as equals as well as to make their dialogue a success (Creswell, 2008). This is a way for accomplishing both a sense of community and a living democracy that action research espouses, via dialogical conversations (Kindon *et al.*, 2010). The dialogue is considered a multi-voice powerful representation, where the voices and responses of others can occur in a non-threatening setting and all this contributed to participants' professional development. The researcher generated data by recording the entire conversations during the meetings, which were later transcribed and analysed. (Turesky and Gallagher, 2011).

3.10 DATA ANALYSIS AND INTERPRETATION

Data analysis is viewed as the statistical or other methods used to process collected data and transformation it into an answer to the original research question while data interpretation means to assign significance or coherent meaning to the collected data (Neuman, 2011). For this study, data analysis and interpretation was done in accordance with the principles of the grounded theory approach. In grounded theory methods, there is no preconceived theory or coding scheme instead an attempt is made to let the data speak for itself, by using the strategies of reviewing, discussing, coding and perhaps model or theory building (Neuman, 2011). Its methods provide a systematic procedure for shaping and handling rich qualitative material (Collins, 1999). Grounded theory methods were relevant for this study and were used because of the quality that they brought to the research. In addition, they enabled the researcher to simultaneously get involved in data generation and to analyse the phases of the research, which was deemed important in the cyclical approach of action research as in this study (Kindon *et al.*, 2010).

The researcher continually interpreted the data by guiding the actions which created analytical codes and categories from the gathered data (Charmaz, 2006). Grounded theory methods were also helpful in developing middle-range theories to explain the behaviour and processes that were needed to improve the achievements of targets for meeting deliverables, which led to the enhancement of the project success theory. Data analysis also involved data reading and organisation through the use of analytical memo writing, coding and categorising as well as enhancing analysis by taking categorised themes to the participants for validation and for further development of strategies that had solved problems which hinder the ability to complete project activities on time (Neuman, 2011). See Figure 3.6 below.

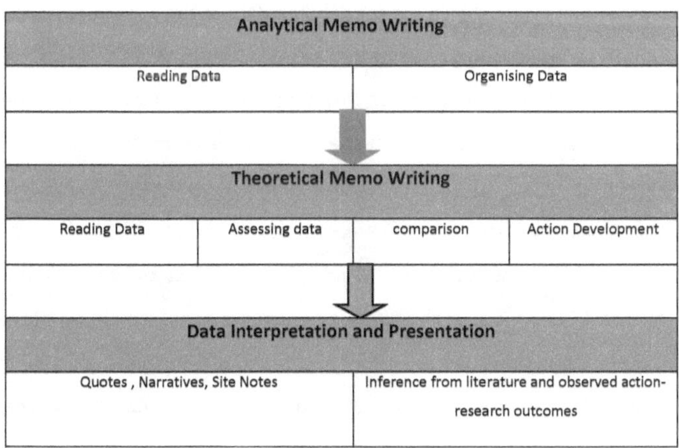

Figure 3.6: Data analysis, interpretation and presentation techniques
(Researchers concept)

In terms of data analysis, the researcher looked for key aspects that hindered the completion of project tasks on time by role-players (Stringer, 2008; Creswell, 2013). The researcher gathered and analysed data on a continuous basis and simultaneously generated data and analysed it as a means that the emerging analysis had shaped the data-generation procedures (Charmaz, 2006; Bryant and Charmaz, 2007; Birks and Mills, 2015). This meant that after every interaction with role-players, the researcher went through the information captured and analysed it. This involved categorising issues into themes and taking those themes to the role-players for validation and further development of strategies that solved problems which hindered their ability to complete respective activities on time, cost and quality (Oliva and Pawlas, 2007; Buys, 2010).

The fact that research involved both the researcher and the participants in the research process was viewed as a collaborative effort, where everyone was involved in the resolution of a collective difficulty (Oliva and Pawlas, 2007). During data analysis, too, it was necessary to relate any knowledge that had been generated to the general social and historical situation and to try to appreciate how the participants had effected and influenced change. Before analysing the data, the researcher first organised it because the organisation of data is important in qualitative research in view of the large amount of information that was gathered during this study (Stringer, 2008; Creswell, 2013). The study concluded that the organisation of data in qualitative research meant keeping data in the chronological order of the events, having a detailed discussion of several themes with sub-themes, having specific illustrations, and having multiple perspectives from individuals, as well as quotations (Stringer, 2008; Creswell, 2013). As a result of the nature of the information involved during the phase of data gathering for this study, the researcher was not obliged to use a computer but instead typed the necessary forms which were completed or filled manually and stored for analysis. Some of the manuscripts were scanned and stored on the computer as electronic copies for back-up. Accordingly, data was arranged according to the source obtained from the interactions, observations, the reflections and the site notes or memos which were safely stored. This required the application of analytical memo writing where the researcher read data or information continuously in order to organise it. Memorandum writing was construed to be essential to grounded theory hence the researchers option to employ it for the current study (Bryant and Charmaz, 2007).

A memo is a unique research tool which helps the researcher to explore what is going on at the research site and memos also conceptualise the data in narrative form. Remaining firmly in the data, researchers used memos to create social reality by discursively organising and interpreting the social worlds of the participants (Bryant and Charmaz, 2007). During the researchers' interaction on the project site, memos of different aspects were written, some of which were about the state of discipline and order during workshop time and the information was compared with that which was obtained from the participants' reflections. The information contained in the researchers' memos was helpful for data analysis and continuous reading of all the data helped to obtain a general sense of the information and to reflect on its overall meaning (Creswell, 2008). Reading all the data to obtain the general sense meant reading the transcripts several times, with the aim of immersing the researcher in the details, in order to get a sense of the data as a whole, before breaking the data into parts (Delport *et al.*, 2015). Breaking down the data into categories involved the application of theoretical memo writing and also required the use of the qualitative data analysis.

An analysis of the data generated was presented by the researcher in the form of lengthy narratives in chapters 4 and 5. This was done in sufficient detail for the purpose of allowing the reader to judge the accuracy of the analysis (McMillan and Schumacher, 2010). In the current study, the researcher also used raw data to illustrate and substantiate the interpretations. The interpretations were substantiated by quoting participants and memos presented by focus group workshops during the sessions (McMillan and Schumacher, 2010). Interpretation of knowledge in qualitative research was understood through making links, interpreting contexts and perceiving meanings attached to the data (Creswell, 2008). Understanding of knowledge was determined from the findings produced by the research, which were interpreted differently at different times and in different places by different people (Zuber-Skerritt, 2011). This called for the use of a number of strategies to ensure the validity of the research design, because the validity of a qualitative research design involves ethics (Stringer, 2008; Creswell, 2013).

3.11 RESEARCH METHODS

Research methods are sets of specific techniques for selecting cases, measuring and observing social aspects of life, gathering and refining data, analysing data and reporting on the results (Neuman, 2011) The qualitative researcher is likely to generate, analyse, and interpret data simultaneously, going back and forth between research steps, because research is an interactive process, in which steps blend into each other (Neuman, 2011). The research methods employed in this study are presented in the following discussion starting with the research site.

3.11.1 Research site

The research site was a school construction project within the Ngaka Modiri Molema District Municipality in Delareyville, South Africa involving the construction of high school. The choice of this site was due to the fact that the researcher's role on the project is that of project manager, which will accord him with the authority to make changes related to what is required of this study. The project which comprised the construction of ten blocks of classrooms, an administration block, kitchen, dining facility, library, computer and science laboratories, bulk sewer system and sporting facilities was faced with so many challenges where the client was contemplating to evoke the penalty clause by imposing penalties before terminating the contract.

3.11.2 Sampling

Purposive sampling seeks to ensure that the diverse perspectives of people who are likely to affect the problem under investigation are included in the study (Neuman, 2011). Purposive sampling seeks to select participants for a variety of purposes, namely maximum variation sampling (which seeks to include people who represent diverse perspectives in any social context), extreme case sampling (which strives to include particularly troublesome or enlightening cases), typical sampling (which endeavours to include participants who are typical of people in the setting), or theory or concept sampling (which tries to include participants who have particular knowledge related to the issue studied) (Stringer, 2008). For the purpose of this study, purposive sampling was not done because of the limited number of participants available for the study. The researcher found it necessary to mention it so that the reader is aware about this type of sampling.

Typical sampling was instead used to select and isolate activities identified as the prime causers of delay for the project as both, the stakeholders and the setting were most typical of this study's population (Wellington and Szczerbinski, 2007; Creswell, 2013).

3.11.3 Ethical considerations

The intention was to work with role-players responsible for the activities cited as problematic in order to devise strategies to help in ensuring that the risks or problems affecting such activities were identified and managed while the process and outcomes are recorded and documented for further analysis. Participants cannot be forced to take part in a study if they are not willing to participate (Neuman, 2011). This led the researcher to seek general permission from the client, who communicated with various firms represented on the project to allow their agents to cooperate with the researcher. Finally, notwithstanding the foregoing, the researcher undertook to get a buy-in from the respective individuals attached to the project in order to gain study. Different letters written to the various stakeholders are role players are presented as Annexures C, being the default letter, Annexure F, being a letter of motivation by the Principal Agent (PA) and Annexure E, being the application letter for extension of time. Resulting from the above, all stakeholders such as the client, employer, professional team, contractor, sun-contractors and community indicated their willingness to work with the researcher.

3.11.4 Data generation for this study

Data generation and analysis in action research is effective when it is accomplished as an interactive process between stakeholders (Kindon *et al.*, 2010; Creswell, 2013). Most of the data for this study was therefore generated from the interaction between the participants and the researcher and from their interaction with one another. Data generation refers to the various ways in which data will be obtained for a study (Neuman, 2011).

3.11.5 Recording of stimulus to project activities (Used during cycle 1 and 2)

Action research is a participatory activity in a specific area of interest facilitated by a researcher (Zuber-Skerritt, 2011; Zuber-Skerritt and Teare, 2013). The role-players will be engaged in action research activities underpinned by the stimulating action in the different pre-determined areas of concern in order to change the modus operandi, build capacity, continuous improvement of professional practice, mutual respect and commitment to establishing and pursuing goals and working together to achieve them (Zuber-Skerritt and Teare, 2013). During the above process, the role-players will be assisted to improve their performance and achieve their goals (Buys, 2010). The researcher, being the facilitator, will act as a coach to facilitate the implementation of the stimulating action to project activities with the aim of achieving improved outcomes. (Zuber-Skerritt and Teare, 2013). In terms of the above, coaching will be focused on helping team members to improve their problems identification abilities and to develop and support them in the overall improvement of their current roles and for future functioning as this is a non-directive form of coaching, focused on helping team members to close the gap between where they needed to be and where they were (Buys, 2010). The participants will require to discuss their current situation, propose their own achievable goals that would effect change and improvement, explore new possibilities that would lead to their desired change, support one another in their plan of action and assess their performance (Buys, 2010).

3.11.6 Credibility and process validity

Credibility means the plausibility and integrity of a research study (Stringer, 2008). Regarding the credibility of this research, the researcher makes it possible for the participants to consciously observe and record events, activities, and contexts over a long period of time. Conscious observation of events and activities means that one has to take notes about what is actually happening at the time, rather than describing what has happened from memory or from an interpretation of what one thinks happened (Neuman, 2011). The notes taken by the researcher were written in a chronological order, with the date, the time, and the type of the action which occurred (Neuman, 2011). In order to enhance credibility of this study, the researcher uses member checking. Participants are given the opportunity to view raw data and conclusions reached for consensus on the stimulating action initiated. This does not only enable the participants to verify that the research adequately represents their perspectives and experiences but also expose participants to raw data and research reports which provide the participants with an opportunity to clarify and add information related to their experience which also helps to prove the authenticity of the research process (Stringer, 2008).

Process validity is a measure which examines the extent to which problems are framed and solved in a manner that permits ongoing learning of the people involved in the research (Herr and Anderson, 2002). Outcome validity is dependent on process validity in the sense that if the process is superficial or flawed, the outcomes will reflect this fact. Process validity must also deal with the much debated problem of what counts as evidence to sustain assertions, as well as the quality of the relationship that the researcher develops with the participants (Herr and Anderson, 2002). To ensure process validity in this study, the project records and trends were reviewed and participants were also engaged in dialectical discussions in several platforms. This required them to apply the strategies that they were agree upon and to reflect on the situation in order to enable ongoing risk management on the part of the participants (May and Govender, 1998).

3.11.7 *Case Study*

The researcher adopted the action research within the single case study domain as a strategy for this current research which is explained in greater detail below;

In order to understand the case study design, it is important to start by establishing what a case study is. Case study is explained as a research strategy that undertakes an in-depth investigation of a single unit (Yin, 2013). For the purpose of the current research, the case study was that of reviewing the current risk management tools and the implementation of project management strategies to manage risk on the construction. A case study is also the development of detailed intensive, knowledge about a single 'case' or a small number of related cases (Sauders *et al.*, 2003). It can also be said that all research types have some form of design which maps out the logical flow which connects the empirical data to the initial research questions which lead to conclusions (Yin, 2013).

While the main objective of the research design was to ensure the collection of data within the parameters of the research purpose to avoid collecting evidence that did not address the research questions, the case study approach was used in this research by virtue of it being an 'action research' in order to try and relay the message using events which occurred, or those that are still taking place (Bell, 2014). Similar to this current research, case studies are particularly appropriate for individual researchers because they give an opportunity for one aspect of the problem to be structured in some depth within a limited time scale (Bell, 2014).

Another view is that a case study could be a site such as an organisation or a department within an organisation and activities could also be viewed as units of analysis in a case study and that a case study could also be a person (Bryman and Cramer, 1990). However, in contrast, the unit of analysis in a case study does not necessarily need to be a site, organisation or human but could also involve both personal, public and official documents such as diaries, minutes, strategic plans, letters and directives (Huysamen, 1994). A case study is further considered to be basically an explanatory piece of research carried out in one field setting by utilising a variety of techniques so that the researcher is not bound by one method but capitalises on any approach that might unravel a new puzzle (Baldridge, 1971). Case studies are reported to be directed at the understanding of uniqueness and idiosyncrasy of a particular case in all its complexity (Huysamen, 1994). There are two aspects which had to be considered when selecting the case study as a preferred option and these are;

a) The case needs to be defined or demarcated so that meaningful boundaries are determined. In this study, the case was clearly defined as a school construction project.

b) Which-ever technique is used to collect data, the concern is directed at what is being observed. The technique used to collect data for this study was the focus group workshops and observation of experimental research through action research in which stimuli introduced to project activities were observed and the outcome recorded and documented.

3.11.6 Action Research

The research style adopted for this study was Action Research (AR) which is defined as a participatory, democratic process concerned with developing practical knowledge in the pursuit of human purposes grounded in a participatory belief which also seeks to bring together action and reflection, theory and practice, in participation with others, in the pursuit of practical solutions (Brydon-Miller *et al.*, 2003). Action research is further defined as an orientation to knowledge creation that arises in a context of practice and requires researchers to work with practitioners (Bradbury-Huang, 2010). Both Brydon-Miller (2003) and Bradbury-Huang (2010) emphasised the fact that action research involved the interpretivist, inductive and qualitative ethnography in which the researcher participated into the subject of the study such as the stimulation of activities to confirm or disprove the attainment of the desired outcomes just like what the researcher has undertaken to do in this current study.

3.11.7 Action research explained

Action research is mostly used in critical social science theory studies as a way to intervene in a problem situation (Huysamen, 1994). Studies previously conducted have revealed that researchers using action research mostly use methods from other paradigms for data gathering and analysis and further concluded that action research is a well-established research method which has been in use for over a century by social and medical sciences (Huysamen, 1994).

3.11.8 The action research approach

Action research was described as a five phase cyclic process depicted in Figure 3.7 (Huysamen, 1994). Initially, a research environment or a client-system infrastructure was established. In this environment, five identifiable phases were repeated and these are diagnosing, action planning, action taking, evaluation and specifying learning. Client-system infrastructure is the specification and agreement that constitutes the research environment. This infrastructure provides the authority under which researchers may specify actions, legitimate actions which will be beneficial to the project and should also define mutual responsibilities between the different stakeholders to each other. A very important aspect was to collaborate the nature of the undertaking whose detail involved diagnosing, action planning, action taking, evaluating and specifying learning (Huysamen, 1994). This phase is usually an on-going process although it is formally undertaken at the end.

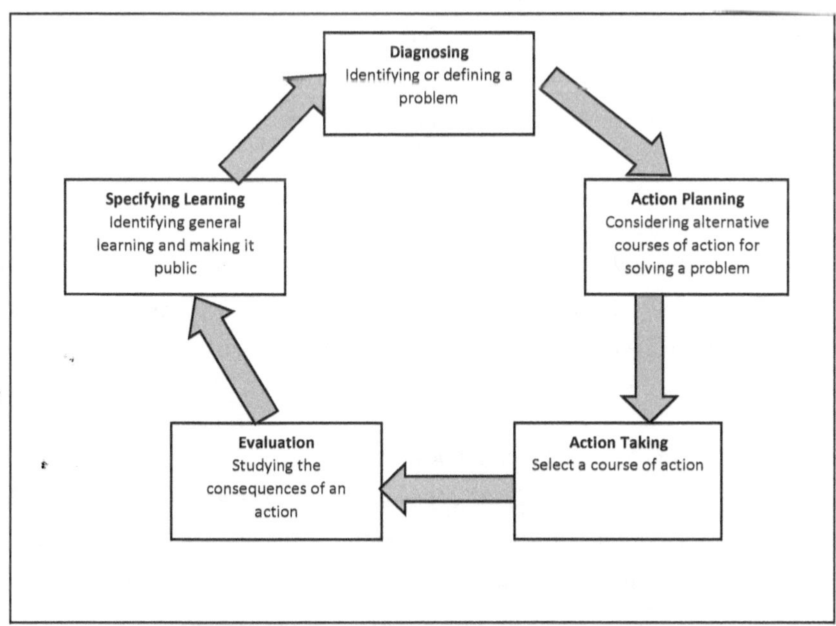

Figure 3.7: Phases within an action research cycle
(Huysamen, 1994)

The research phases within the action research domain depicted in Figure 3.7 were adopted for use by the researcher in this current research. The phases were similar in nature to those phases referred to in chapter 2 which were identified as the orientation phase, information phase, function phase, evaluation phase, development phase, presentation phase, and implementation phase. These are described in detail in paragraph 2.3 and they are the phases which this research adopted for the research site where the researcher was employed as project manager on a school construction project in the Ngaka Modiri Molema District Municipality of the North West Province at which the underlying assumptions of this study were reviewed, observed and results recorded for the purpose of achieving the research objective.

3.12 SAMPLING PROCEDURE AND SIZE

The case study for the current research was the construction project site at which the researcher was employed as project manager. No sampling was done because of the limited number of participants which rendered the sampling process irrelevant. The action-research approach within the case study design was employed which also incorporated an embedded survey within the case study where some project stakeholders relevant for this purpose are participants by means of information gathered during the focus group workshops. The total study size used in this research is 15 participants drawn from project role-players according to Table 3.3.

3.13 STUDY POPULATION

The study population comprised of all role players within the stakeholders of the construction project who included, the professional team, client, employer, implementing agent, project steering committee, contractor and sub-contractors. The total number of participants was 15.

3.14 METHODS OF DATA COLLECTION AND ANALYSIS

In order to collect data from the selected samples, several methods were available to the researcher for this study, who preferred to use the case study procedures for the purpose of this research. The methods of data collection were divided into two approaches being;

a) The behavioural approach where one can observe conditions, behaviour, events, people or processes and (Cooper *et al.*, 2006).

b) Communication approach where one can communicate with people about the various topics (Cooper *et al.*, 2006).

It is argued that one can observe conditions, behaviour, events, people or processes (Behaviour approach) or one can communicate with people about various topics (Communication approach) both of which are presented in the following discussion.

3.14.1 Observation Approach

The observation approach of research qualifies as a scientific enquiry when it is conducted specifically to answers research question one as it was systematically planned and executed using proper controls and provided a reliable and valid account of what happened (Cooper *et al.*, 2006).

3.14.2 Communication Approach

The communication approach involves questioning or surveying people and recording their responses for analysis (Cooper *et al.*, 2006). The great strength of questioning as a primary data collecting technique is that it does not require a visual or other objective perception of the information sought by the researcher as abstract information of all types can be gathered by questioning another (Cooper *et al.*, 2006).

3.14.3 Data collection methods

The data collection methods for this study was the use of memos and observations which are documented for analysis through samples presented as Annexures F and I, being the memorandum and data collection sheet respectively.

3.14.4 Data analysis

Data analysis was done by means of observing the outcomes of the results during cycle 2 investigation following the introduction of stimulus to project activities and project management processes. The analysis of data was further done by comparing the effects of the stimulus on the outcomes before and after the introduction of project management processes. The results were then interpreted and conclusions drawn.

3.15 ETHICAL CONSIDERATIONS

Ethics are an important part of any research especially one which deals with people and animals as subjects of the study (Welman *et al.*, 2005). The Ethical approval for this study was granted by the University of Cape Town through the Ethics Committee and a copy of the approval is presented as Annexure M.

3.16 SUMMARY

The aim of this chapter was to provide the reader with an insight and understanding of the research design in which the methodology and processes adopted for this study were presented in detail. Action research was the main research approach used and this chapter also presented information on the various aspects which related to the three research paradigms referred to as, positivism, interpretivism and critical social theory. This research was further based on the deductive approach and a combination of both the qualitative and quantitative methods. The study employed the combined approach because of the advantages inherent to both methods and the threat of restrictions envisaged upon the use of only one methodology was eliminated. The study also used both the behavioural and communication approaches as this was done for the purpose of not confining the research only one approach which was construed to be restrictive hence the option to use both the observations and narratives as data collection tools. The results of this study are presented in the next chapter.

CHAPTER 4: RESULTS AND DISCUSSION OF FINDINGS

4.1 INTRODUCTION

This chapter presents results obtained from the study undertaken in accordance with the research design which stated that the study was conducted in two phases namely; cycle 1, comprising the data gathered from interviews where the researcher interacted with 3 focus groups participants in workshops and cycle 2 which comprised the action-research component also undertaken within the case study design. The 'cycle 1' results present the demographics and summary of responses sourced from the focus group participants while 'cycle 2' results present the outcomes of the stimulating actions introduced to project activities aimed at invoking respective reactions which are observed, recorded and analysed. The criterea for the development of stimulating action is selected from an analysis of the information gathered from focus groups and the data obtained from literature where views of scholars are analysed to provide possible causes of problems currently faced In the construction industry especially those affecting the project success criteria used in this study. The purpose for the two-pronged approach was to ensure the existence of a strong basis to validate the findings to the topic, "A review of risk management tools for construction projects and the implementation of project management strategies" investigated by this current research.

The following section provides an overview of participants' demographic information according to the embedded survey among the focus groups. This is followed by a summary of responses provided by participants in the focus groups. A detailed discussion of cycles 1 and 2 is presented in the following section.

4.2 CYCLE 1 INVESTIGATION

4.2.1 Survey within workshop participants

A total of 15 people drawn from project stakeholders were selected as participants and placed in 3 groups, each consisting of 5 participants for the workshops facilitated by the researcher. The criterea for selection was based on the role of the respective stakeholders on the project, the link between their roles and how their focus areas were associated with the presumed causes of challenges on the construction project site as highlighted in the preceding chapters 1 and 2. The biographical information of the focus group participants was presented in Table 3.3 in the previous chapter. As stated above, a total of 15 participants from different disciplines within the construction sector assigned to the project participated as respondents in the embedded survey within a case study design conducted by means of focus group workshops. The participants were drawn from a cross section of project stakeholders described in section 4.2.1 who were also representatives of consulting firms within the built environment contracted for different roles on the project including the client, the employer, the professional team (consultants) and the contractor and sub-contractors. The names of participants and those of consulting firms were withheld in order to comply with the anonymity undertaking made by the researcher to the Ethics in Research (EiR) Committee of the University of Cape Town (UCT).

In compliance with the anonymity undertaking, the researcher coded participants who were referred to as 'Respondent 1' to 'Respondent 15' denoted by 'R1', 'R2', 'R3' to 'R15' in this study while all juristic persons were also coded and referred to as Company 1, 2 denoted by 'C1', 'C2' to 'C15'. This was done to ensure that no response or inference was traced back to any of the respondents or firms which participated in this study.

4.2.2 *Data generation and analysis in cycle 1 investigation*
Most of the data for this study were generated from the interaction between the participants and the researcher as well as from their interaction with one another. Data generation refers to the various ways in which data is obtained for a study (Welman *et al.*, 2005). In qualitative research, there is a wide range of methods of data generation which includes, personal experience, life stories, interviews, observations, reflections and interactions (Welman *et al.*, 2005). Data for cycle 1 of this study was generated by way of interactions between the researcher and participants using workshops where the researcher utilised the qualitative approach to collect data through the personal experiences and reflections of participants on project operations.

The process for Cycle 1 investigations was conducted through workshops facilitated by the researcher with each focus group at different times. During the workshops conducted at the project site, the researcher started by presenting an orientation of the workshop to participants and explained the status of the project at the time, where it was emphasised that the project was behind schedule as most project activities had not been completed on time and were still lagging behind. Participants were requested by the researcher to freely discuss and identify what they viewed as the main causes of delays which had ultimately led the project to lag behind schedule. Participant were each given a memo on which to write a list of five reasons considered to be the causes of delay. A copy of the memo is presented as Annexure F and the summarised responses from focused groups are presented below.

4.2.3   Focus group 1 response

Data collected from focus group 1 participants indicated that the main causes of project delay were amongst others as; late ordering and delivery of material by the contractor, constant contract scope changes by the client, design related changes by the engineers, delays in responding to Request for Information (RFI), cashflow problems resulting from delays in processing of payment by the client, the contractor's inability to properly sequence project tasks and interference from the community. Some of the responses from focus group 1 resonated with conclusions drawn from literature review by scholars who asserted that risk on construction projects was compounded by the involvement of many role players such as the client, stakeholders, contractors and designers (Welman *et al.*, 2005). It is also evident from the analysis that the researcher conceptualised the actions which comprised the stimuli introduced to project activities during cycle 2 investigations such as 'Action 2 which related to the implementation of a procurement plan for materials, plant, tools and sub-contractors. The above response also gave rise to the conceptualisation of 'Action 4', which referred to the implementation of a stakeholder management plan. Stakeholder management plan finds prominence within the project management knowledge areas, where it places an emphasis on the identification of customers' needs, which is a prerequisite in a project charter because ignoring this component directly results in the identified problem of constant scope changes by the client (PMI, 2015). Action 2 was also important as part of management strategy because the process identified influential stakeholders among a cross-section of stakeholders and clarified their needs in order to manage their expectations from the project (PMI, 2015).

4.2.4   Focus group 2 response

Focus group 2 was comprised of the main contractor, sub-contractors, civil engineers, electrical engineer and representatives from the principal agent's office. Data gathered from focus group 2 indicated that the causes of project delays were wrong material orders, late placement of orders for material, inability for the contractor to buy material in bulk, site closures or work stoppages due to civil strife by the local community, inability to understand the project requirements, lack of full project knowledge, poor workmanship leading to demolitions and reworks, additional works, lack of punitive action for non-compliance, lack of proper monitoring and control of tasks and project activities and enlisting incompetent people to execute tasks and activities. The information gathered from focus group 2 was also supported by conclusions revealed in the literature review where it was reported that lack of financial and management skills, lack of experience and capital to finance projects by black contractors were some of the causes of the construction related problems, which were also referred to as industry risk (Jansen and Christie, 1999; McGregor *et al.*, 2017). It was on this basis that stimulating actions 2, 3, 4 and 6 were conceptualised for cycle 2 investigations. Action 2 was informed by the assertion that delays were caused by late or wrong material orders, action 3 was informed by the assertion that poor workmanship which led to demolitions and re-works was a result of lack of competence and skills hence the need to conduct a skills audit in which trades and competencies were verified.

Action 4 was initiated as a result of the conclusion that site closures were due to protest action and civil strife by community members which caused project delays hence the need to implement a stakeholder management plan in which the needs and expectations of the community, being an important stakeholder were identified and managed (Jansen and Christie, 1999; McGregor *et al.*, 2017). Action 6 resulted from the assertion by focus group 3 participants that lack of punitive action for non-compliance caused sub-contractors to neglect their responsibility to comply with the set standards to attain quality led to many demolitions of sub-standard work. This resulted from lack of the enforcement of punitive measures which would serve as a deterrent hence the need for 'action 6' which emphasised the enforcement of the penalty clauses within the contract under the general theme of contract management.

4.2.5   *Focus group 3 response*

The third workshop comprised focus group 3 made of participants drawn from the Community liaison officer (CLO), Social facilitator, Sub-contractors for tiles, bricks and roof respectively. Data gathered from this interaction with focus group 3 participants revealed that problems affecting the project were, lack of communication between the community and the contractor which led to community protests on site, late payments, low payment rates to sub-contractors, adverse weather such as rain and extreme cold, self-imposition of sub-contractors for subcontracted project work and conflicting instructions issued by consultants for the same task. Most of the responses of focus group 3 were similar to those presented by focus groups 1 and 2 except conflicting instructions issued by professional team members on the same task. The data generated indicated that the contractor and sub-contractors were given different conflicting instructions by different consultants for a particular task or activity because the different people assigned to the project had different ways of doing things and every new person to the project issued a new and different instruction on how to execute the same task. This caused task implementation problems which in some instances caused delays while in other instances led to demolitions. This led to the incorporation of items 1 and 5 as stimulating action. This referred to the signing of a Memorandum of Understanding (MoU) and Team Induction respectively in which the MoU was signed with respective role-players to minimise the high staff turnover of people assigned to the project and where such was unavoidable, then the Induction of team members was implemented in order to appraise the in-coming new team member on the activities, plan and any other vital information regarding the task.

4.2.6 *Analysis of Cycle 1 Data*

The data from responses provided by focus group participants were analysed in terms of similarities across the groups and compared with the views presented in the literature review on the possible causes of delayed construction projects within the project success criterea constraints of time, cost and quality. The literature reviewed for this study revealed that the challenges which have led to the slow pace were caused by among others, delays, cashflow problems, lack of financial management skills and experience to manage projects, appropriate competences, other construction related skills and poor quality work (Jansen and Christie, 1999; McGregor *et al.*, 2017). It also revealed that project success criterea was mainly measured in terms of time, cost and quality because an alteration to one of these constraints has an effect on the other as the essence of value management is to eliminate anything that adds cost essential to the basic function of the product (Jansen and Christie, 1999; McGregor *et al.*, 2017). The researcher assessed the above conclusions and made comparisons for similarities with data generated from focus group workshops which led to the development of six general areas employed in the action research part of the case study which comprised cycle 2 investigation.

4.3   CYCLE 2 INVESTIGATION: Action Research

The second part of the study is cycle 2 investigation done through Action Research (AR) in which some stimulating action was introduced to selected project activities in order to observe the outcome. The stimuli could be introduced to different other activities but the selected activities were preferred from the many activities because of the close causal link with what was ascribed to as the possible reasons for project challenges through literature review and data generated from focus groups workshops.   The stimulating actions were introduced to project activities through the use of processes adapted from the various value management protocols highlighted in chapters 2 and 3 respectively. The employed value management protocols were derived from standardised and documented project management strategies (Jansen and Christie, 1999; McGregor *et al.*, 2017). The above approach was aimed at evaluating the implementation of the above project management strategy as an effective tool for management of risk in the forms that have been highlighted in the preceding chapters reported to be the cause of failure by contractors to achieve project success.

4.3.1   Case study description

This research was underpinned by action research premised on a case study conducted on the construction project at which the researcher was a Project Manager. This was a building project involving the construction of a school in the Delareyville area of the Ngaka Modiri Molema District Municipality in the North West Province of South Africa. The initial contract duration for completion of the works is 16 months and the scope of works include the construction of ten blocks of classrooms with each block containing four classrooms, an administration block, a multi-purpose kitchen and dining hall, two separate ablution blocks for boys and girls respectively, a science laboratory, a computer laboratory, a media centre, sport facilities which included a combi court (Basketball, tennis and volleyball courts), a soccer pitch and a netball court and also a bulk septic tank equipped with the biosorp system. The initial total project value was R35m but the total later escalated to R45.5m after the application for extension of time was granted at an additional cost of R10.5m.

The status of the project at the time of the case study, in March 2018 was that the project schedule had been exceeded by more than ten months. The work progress was at 30% while 60% of the allocated budget had already been spent. At that time, the project was behind schedule, and projected to have an over expenditure on the budget of approximately R10m. The PA's recommendation to the client to institute penalties against the contractor placed the contractor in a bad financial position in terms of cashflow because at that time, the PA had already advised the client to initiate termination procedures so that the contract is terminated. The effect of contract termination is the submission of the contractor's particulars to the National Treasury for the purpose of blacklisting. The consequence of blacklisting any service provider to government is that the name of the service provider is entered on the database of blacklisted suppliers by the Treasury and the affected supplier is banned from procurement of government services for a period of five years.

For the purpose of this study, the researcher in his capacity as project manager submitted an application for extension of time, presented as Annexure E, which was approved through the approval letter presented as Annexure G and the project duration was extended by twelve months from the date of approval with cost. The approval with cost meant that additional funds equivalent to the outstanding project activities were allocated for the completion of the project. The researcher then undertook to reorganise the project in terms of the research design in order to reduce the various challenges on the project which had translated into project risks by the use of action research within the value management domain. The starting point was the careful selection of stimulus introduced to project activities. It is also worth noting that the delineated activities were selected on the level of importance and their impact to project success by the inferences from focus group data and conclusions drawn from literature review.

The process of introducing stimuli to project activities was done by using the phases within the value management cycle implemented as part of the project management strategies to improve project success. It should also be noted that there were many other activities available to the researcher which would have been used but the researcher opted to limit himself to the activities in Table 4.1 for the purpose of restricting the activities to a manageable scope. A detailed description of actions taken for the cycle 2 investigation is presented in Table 4.1, and explained.

ACTION	STIMULUS	PURPOSE
1	Signing of a Memorandum of Understanding (MoU) with the professional team	manage risks associated with high staff turnover according to the focus group responses and literature review (Thomas, 2013; Bilau, 2015)
2	Implementation of a procurement plan for materials, plant, tools and sub-contractors	Identify and manage procurement related risks according to the results from focus group workshops
3	Skills audit, trades and competencies	Manage risks related to quality, cost and time according to the results from focus group workshops and literature review (Baloyi and Bekker, 2011; McGregor *et al.*, 2017)
4	Implementation of a stakeholder management plan	Manage risks related to stakeholders (Al-Bahar and Crandall, 1990; Baloi and Price, 2003; PMI, 2015)
5	Team Induction	Manage time and conflict related risks according to the results from focus group workshops and literature review (Thomas, 2013; Bilau, 2015)
6	Contract Management	Enhance project accountability. Data sourced from focus group workshop participants and literature review (Ngwenya, 2007; Baloyi, 2011)

TABLE 4.1: Stimulating action introduced to project activities

(Researchers concept)

4.3.2 *Action 1: Signing of the MoU with professional team*

The various forms of action depicted in Table 4.1 illustrate the details of stimuli introduced to project activities in order to observe and record the results. The first stimulating action introduced was the signing of a Memorandum of Understanding with role-players.

During this phase, a MoU was signed with all the companies sub-contracted to provide various professional services to the project through their respective representatives such as the Architect, Quantity surveyor, Civil engineer, Structural engineer, Electrical engineer, Social facilitator and the Occupational health and safety (OHS) consultant who were a part of the professional team. A sample of the MoU is presented as Annexure J. This process was conducted in accordance with the phases within the action research cycle of diagnosing, action taking, evaluation and specifying learning based on Figure 3.6 in the previous chapter

3. The process conducted within the action research cycle was also conducted during Action 1 with the purpose of ensuring that the delays caused during orientation of new team members and differences in their approach for executing certain activities were managed. This was done in view of data analysed from focus group 1 workshop participants and according to conclusions reached by some scholars in the literature review (Thomas, 2013; Bilau, 2015).

4.3.2.1 Diagnosing

Diagnosis is described as a judgement about what a particular problem is (McIntosh, 2013). The underlying problem related to the lack of stability caused by a high staff turnover within the professional team members assigned to the project by the respective consulting firms sub-contracted to the project for various services. This was one of the identified potential risks reflected in the previous chapters as well as an inference made from the information provided by the focus group during the workshop. The risk was mainly based on the delays resulting from the newly assigned people to the project because so much time was spent on orientating the new team members to both the project and other related project role-players. This also affected the newly assigned person's ability to move at the same pace on project activities as other existing members of the professional team as revealed in the literature review. Also identified as a potential problem was the difference in the level of experience and competence between the predecessor and the successor (Bilau, 2015).

The above risk was emphasised in the fact that the person withdrawn from the project was more experienced and competent than the new person brought onto the team, which in certain situations resulted in the difference of approaches adopted to handle project activities (Thomas, 2013). The other risk diagnosed related to the constant changes in design which had an effect on both the quality and time constraints of the project. The results revealed that errors in designs and drawings also had an impact especially when the newly assigned team member to the project relied on the design which had an error as this caused further time delays to correct through the issuing of revised drawings (Goede *et al.*, 2012).

4.3.2.2 Action Planning

During this phase, a plan to manage the risk of high staff turnover on the project was conceptualised with the aim of finding a way to ensure that the high staff turnover on professional team members was significantly reduced. This plan included obtaining of a commitment from firms sub-contracted to the project to make sure that unnecessary replacement of professional team members on the project was minimised and where it was unavoidable, proper team induction of newly assigned team members was complied with in accordance with action number 5 in Figure 3.6.

4.3.2.3 Action Taking

A Memorandum of Understanding (MoU) is signed with each of the consulting firms, forming the professional team on the project where an undertaking was made by all firms to retain and maintain their staff who were part of the professional team of consultants assigned to the project for the entire duration of the project without being replaced unless the situation was unavoidable.

4.3.2.4 Evaluation

The action taken during this phase was evaluated and the foremost observation was that turnover on the professional team members decreased drastically from an average of three every month to zero. The consistency in the team members further contributed to the stability on the construction project site. This is done in accordance with what the literature review revealed relating to the success of the intervention which was said to be the most important in any critical research project (Goede *et al.*, 2012). It was also noted that design related disputes and general strategic approach disputes among the professional team were no longer experienced on the project. This was because of consistency in the agreed upon project execution protocols which were planned during the initial technical meeting before the implementation of the conceptualised actions for the duration of the project. Time spent on orientation and induction of new team members which previously accounted for time wastage was also reduced and ceased to be a cause for concern.

4.3.2.5 Specified Learning

The outcome of the action introduced in paragraph 4.3.1.1 drew the researcher to conclude that high staff turnover in the members of the professional team members had an effect on the constraint of time and quality on a construction project. The reasoning is owed to the fact that improved quality and less time wasted were observed when Action 1 stimulus was introduced to the project activities. Stability in the professional team on the project also enhanced a good working relationship between team members as no member complained to the project manager against a fellow team member for either unnecessary delays, time wastage or design and frequent task sequence changes. The specified learning resulting from the signing of an understanding was that a stable and consistent professional team resulted in good working relationships among team members and prevented unnecessary delays. This also led to the maintenance and compliance with set project quality standards.

4.3.3 Action 2: Procurement Plan Implementation

The following was undertaken during this project phase;

4.3.3.1 Diagnosing

The diagnosis related to the risks associated with the three project constraints of time, quality and cost. This action was derived from the data obtained from the focus group participants who indicated that material related risks such as wrong orders, delays in the ordering of material and the problems of exchanging the wrong material for the correct material caused so much delay and this had an effect on time and cost constraints. it was also noted that one of the risks which had an impact on the time constraint was the late delivery or non-delivery of materials to site. Also diagnosed as a potential risk was that some sub-contractors did not possess the required competencies and experience to execute certain project tasks or activities. This was because their employees either lacked the required skills, expertise and experience demanded by the project for the respective activities. This was in accordance with the information gathered from the workshop with focus groups and also data analysed from literature review (Thomas, 2013; Bilau, 2015).

4.3.3.2 Action Planning

The action plan involved a strategy aimed at ensuring that:

a)	Procurement for all sub-contracted services was done through closed tender from sub-contractors who met the minimum conditions set-out in the contract document and supported by various statutory certification such as Cidb grading, NHBRC and the desired experience verified by at least three references.

b)	Reviewing and developing a material schedule log which specifies the type of material to be ordered, quantity, minimum stock level to order and current stock quantities.

c)	Verifying qualifications, skills, competencies and experience submitted by every sub-contractor' labour on the project.

4.3.3.3 Action Taking

During this phase, the researcher undertook the following activities;

a) Technical specifications in the main tender document were reviewed and summarised in point form by the researcher in his capacity as project manager. This was followed by an audit of the main contractor as well as all sub-contractors appointed to the project for various work such as electrical, plumbing, brick-laying, roofing, paving, steel fabrication, ironmongery and carpentry to ensure that the main contractor and all sub-contractors appointed for the various trades were compliant with the provisions of the tender(contract) document. The parties who were verified to be compliant were approved and instructed by the project manager to continue with the provision of their services to the project while those who were non-compliant were given 14 days to correct their respective defects and comply with the provisions prescribed in the tender(Contract) document.

   The audit by the project manager revealed that, three out of the eight brick-laying sub-contractors were non-compliant one out of the three paving sub-contractors were non-compliant and two out of the five roofing sub-contractors were non-compliant.

   All the six non-compliant sub-contractors were given 14 days to remedy their compliance defects. Seven days after the date of the notice, five sub-contractors presented their compliance documents to the project manager, who reinstated the sub-contractors with immediate effect upon the successful assessment of the compliance documents re-submitted.

b) An audit of all materials in stock as well as all the material requirements for the following three months was conducted. This process involved the proper recording of codes for the different material as specified in the Bill of Quantities (BoQ). Potential suppliers for materials were identified and three quotations for every material item were sourced and compared for price, quality, availability and lead delivery time. The suppliers who offered the most convenient service at a competitive rate were selected and shortlisted. Orders with a lead time of between 20 and 30 days were immediately placed. Coded storage and a Site Materials Accounting System (SMAS) were designed, developed and put in place.

c)   All qualifications, skills and experiences of the main contractor, sub-contractors and their respective employees were reviewed and verified through a referencing system. The institutions which issued the certificates were contacted to verify the authenticity of the various certifications held by respective employees of the main contractor and sub-contractors. All former employers cited by employees were also contacted to verify the period of service, work attributes and competencies including work experience. This was also done in all sub-contractor firms to verify the information supplied by both the sub-contractors and their respective employees about places where certain experiences supplied in the contract documents were gained.

4.3.3.4 Evaluation

The researcher evaluated the outcome of action 4.3.3.3 and noted that the intervention had a positive effect on the project and in the process, improved quality, saved time and also saved funds through the prevention of placing orders for wrong materials as well as time wasted on returns of wrong material for replacement with correct material. This was only in cases where the material was not custom made such as steel which was ordered and cut according to the specifications provided by the contractor. In such cases, it was considered as a waste of funds as such expenditure was regarded as a sunk cost which was not recoverable and accounted for an increase in the project budget.

4.3.3.5 Analysis and Specified Learning

The lesson drawn from the implementation of the procurement plan indicated that it was important to put in place all measures prescribed in the tender (contract) document on basic skills requirements for the project labour force. The action implemented also highlighted the need for contract management to be complied with on the project especially the strict adherence to auditing of qualifications, skills and experience. Interventions on the procurement of sub-contractors and material logistics was also identified as an area which required effective management to ensure that the procurement related risks were managed and prevented from occurring.

4.3.4   Action 3: Audit of qualifications, skills and experience

The process involved in this phase was done in accordance with the process complied with in the audit of qualifications, certifications, skills, competencies and abilities as discussed in sub-sub section 4.3.3.3. This action was informed by the views expressed in the literature review and preceding chapters which concluded that Bantu education deprived Black contractors of the ability to acquire appropriate skills to effectively manage construction activities which would enable such contractors to achieve project success (Ngwenya, 2007; Baloyi, 2011; McGregor, 2017).

4.3.5   Action 4: Implementation of a Stakeholder Management Plan

The following activities were carried out during the introduction of Action 4, which involved the implementation of a Stakeholder Management Plan;

4.3.5.1 Diagnosis

Diagnosis for the purpose of this study is described as judgement about what a particular problem is and in the case of action 4, it is the judgement about challenges inherent with the lack of stakeholder management which have an effect on project success. The diagnosis relates to the risks associated with the three project constraints of time, quality and cost. The challenges associated with lack of stakeholder management as diagnosed in this study were those which mainly involved community protests, delivering a project which the intended end-user was unable to use or did not need and in certain cases, delivering a project not in accordance with the expectations of the stakeholders. This was mainly due to the inability to manage stakeholders in which all stakeholders needed to be identified, their requirements noted and expectations managed. This phase also included the identification of influential stakeholders in accordance with the Project Management Body of Knowledge (PMI, 2015). The conceptualisation of action 4 was derived from data generated from focus groups and literature review in which previous research concluded that some projects could not be used by the user clients or customers because such projects were not compatible with the intended purpose resulting from the inability to conduct a proper stakeholder management process (Al-Bahar and Crandall, 1990; Baloi and Price, 2003; PMI, 2015; Musakwa, 2014; Goetz and Schaeffler, 2015; Naidoo, 2013; Chain, 2009; Preuss *et al.*, 2014; Reid, 2006; Beavon, 2001).

4.3.5.2 Action Planning

An action plan aimed at mitigating the risks pertinent with stake holder management from happening were put in place. This was done through the implementation of a stakeholder management plan in which the key stakeholders were engaged and their expectations identified and managed.

4.3.5.3 Action Taking

During the action taking phase, the following activities were conducted;

a) Community liaison and engagements

The communities within the project site vicinity were engaged through liaison. It was during these engagements that the community indicated what it expected from the project such as:

- The need to give first priority to community members for all casual jobs and general labour work on the project which did not require any formal qualifications, skill or experience.

- The need to appoint a Project Steering Committee (PSC) from within the community to represent the concerns of the community on the project.

- The need for a Community Liaison Officer (CLO) to be appointed from the local community and employed by the project as a link between the project (Contractor) and the community. The role of a CLO was to mediate between the contractor and the interests of the local community. The role also included prioritising members of the community for any job or sub-contract opportunities on the project. The role of the CLO was to further ensure that disputes between the workers and the contractor were resolved timeously before escalating into unmanageable situations.

b) Identification and management of clients' expectations

Several interactions were made with the client in which the expectations of the client were identified and managed. The issues which arose from this process were:

- Clarification of the identity of the project and confirmation that the project envisaged by the client was the construction of a school as specified in the tender document.

- Clarification and confirmation that the desired sports field were the combi-court, soccer pitch and basketball court.

- Clarification that the project was the construction of a co-education school with separate toilet facilities for boys and girls.

- Emphasis was also placed on the need for a library, ten blocks of 4 classrooms each, a media centre, a science laboratory, a kitchen linked to the dining hall and an administration block which comprised offices, a staff room, reception area, a strong room, a nursing room for those sick people, a kitchen and separate toilet facilities for male and female teachers.

- Influential project stakeholders were also identified and managed accordingly with regard to the individual influence each one of them exerted on the project respectively.

4.3.5.4 Evaluation

The process of stakeholder management played a critical role in ensuring that all stakeholder requirements and expectations were addressed and the process further ensured that stakeholders remained content with the expected project outcomes.

4.3.5.5 Analysis and Specified Learning

The lesson drawn from action 4.3.5, which was stakeholder management revealed that the community no longer staged protest action against the project or contractor supposedly for being left out on employment and other opportunities. Continuous dialogue further ensured that the community understood the dynamics of the project and also appreciated the fact that the finished product was for the benefit of the children from the community. The school which the children were attending at the time was far away, dilapidated and neither fit for human habitation nor learning conditions.

Meeting the clients' expectations also proved beneficial as the user client did not at any time direct the Implementing Agent (IA) to stop the project due to the envisaged end product being different from what was expected. Instead, the project was progressed in accordance with the expected and desired outcomes of the client as a result of which the risks inherent with stakeholder dissatisfaction such as delivery of a wrong product not desired by the client or a product which did not comply with the required specifications of the end-user were identified and managed through this action.

4.3.6   Action 5: Team Induction

When the action 5 stimulus of team induction was introduced to the project activities, the following processes were carried out;

4.3.6.1 Diagnosis

The risk of directing efforts into different directions at the expense of achieving 'Unity of Purpose' was one aspect which led to the implementation of this action 5. The reason for the implementation of the 'team induction' action was informed by the data generated from focus group workshops and conclusions derived from the various literature reviewed (Thomas, 2013; Bilau, 2015). This was because it was observed that different team members and sub-contractors joined the project at different times and during different phases of the project. Some of the new role players appeared ignorant of what was expected of them and this resulted In the performance of tasks which were reserved for execution during a later phase of the project. This was due to the fact that the tasks referred to were not sequenced to be executed at the time when certain activities were being performed.

4.3.6.2 Action Planning

A plan was put in place for every new person or sub-contractor working on the project for the first time to be inducted, provided with training or orientation of the project activities pertinent to the particular phase within the respective area of operation in order to appraise the person or sub-contractor on the activities within that phase. This was also aimed at ensuring that the new project staff was made aware of how their respective roles fit into the overall project vision and also advised on the expected time frames for the delivery of various milestones so that the project remained on schedule.

4.3.6.3 Action Taking

During this phase, only the roofing sub-contractor (C8), Sports fields sub-contractor(C11), concrete works sub-contractor (C14) and a few labourers (General workers) were taken through the induction sessions when they reported to the project site for the first time.

4.3.6.4 Evaluation

The induction and orientation sessions resulted in a seamless continuation of work as activities performed by different teams were co-ordinated leading to a stable work-flow according to the programme. Action 5 further ensured that no activity was stopped on account that the preceding task had either not been performed or had not been completed in time resulting from a role players' lack of knowledge as to when such an activity was to be complete or how soon to complete the preceding activity and also identifying which activities needed to be executed simultaneously.

4.3.7   Action 6: Contract Management

The following processes were undertaken during the introduction of action 6 stimulus to project activities;

4.3.7.1 Diagnosis

The action was taken in accordance with the views derived from the literature review in which scholars asserted that one of the major risks causing the school infrastructure backlog was the premature termination of contracts by the client or abandoning of the projects by contractors. It was also suggested that the risk emanating from the lack of contract management was caused by the inability of the contract managers to enforce the prescribed contract penalty clauses when it became necessary to do so. It was further concluded that in some instances, the contract management action was taken so late in the contract that the desired outcomes were not achieved (Kaseke, 2011). Lack of monitoring and control of the penalty clause during default in the early stages resulted into multiple defaults which the contractor(s) failed to remedy because they were numerous and too costly.

4.3.7.2 Action Planning

The plan of action conceptualised was to ensure that anyone in default or in breach of the contract was provided with a written notice indicating the breach and given a number of days as provided by the contract to remedy the situation or have the contract penalty clause invoked in case of failure to implement corrective measures aimed at providing a remedy for the breach.

4.3.7.3 Action Taking

During this phase, the main contractor, C1, was issued with a default notice which indicated that the contractor was in breach of the contract because the project was behind schedule. The contractor was advised to take all necessary measures to restore the project to schedule. In response, the contractor applied for extension of time (assisted by the researcher) by means of a letter presented as Annexure E. The Implementing Agent (IA) referred to as C5 wrote a motivation letter to the client recommending the approval in a letter presented as Annexure K. The clients' approval to the contractors' application was presented as Annexure G. The concrete works sub-contractor, C8 was issued with a notice to correct defects on the concrete works through a site instruction issued in the site instructions book by the structural engineer. At the expiry of the 14 days' notice period, remedial work had not been carried out and the contract was accordingly terminated by the researcher in his capacity as project manager. A new sub-contractor sourced from the site database was procured and enlisted to complete the balance of the concrete work which, the newly appointed sub-contractor did in an accelerated and compressed programme within the stipulated time and restored the programme back to the prescribed project schedule.

4.3.7.4 Evaluation

During the evaluation phase of this action, it was noted that if the penalty clause had not been enforced, both C1 and C4 would have continued performing their respective activities at the same pace which would have had negative effects on the project schedule, and also with dire consequence on quality which had the potential to increase cost and cause a budget over-run.

4.3.7.5 Analysis and Specified Learning

The analysis of the above action implies that sub-contractors whose work was below the specified quality were held accountable and compelled to take remedial action to correct their respective mistakes. The enforcement of the penalty clause served as a deterrent and a way of managing default related risks which had an effect on the project constraints of cost, time and quality.

4.4 PROFILE OF THE ACTION RESEARCH EXERCISE

This section presented the way action research was profiled by the researcher in order provide the reader with insight of what transpired. The discussion also took into account both cycles of the investigation in accordance with the main research question and the related sub-questions which are presented in the following discussion:

4.4.1 *Main research question*

Can an Integrated Approach for Value management be used as a project management strategy for risk management on construction projects?

This question was adequately addressed by means of the investigation conducted in this study through cycles 1 and 2. Cycle 1 utilised the project management strategies by employing the value management approach premised on the conclusions of the literature reviewed which proposed the use of both the value management procedures as adapted from the Generic Value Management Process presented in Figure 3.5 and the procedures adapted from the Soft Value Management approach presented at Figure 2.8 (Jansen and Christie, 1999; McGregor *et al.*, 2017). The main research question was also verified by the assertion of another scholar who defined value management through action-research as a structured, facilitated process in which decision-makers, stakeholders, technical specialists and others work collaboratively to bring about value based outcomes in systems, process, products and services (Barton, 2000). In the case of this study, the researcher, being the Project Manager, assumed the role of decision maker at the level of the project construction site while other stakeholders included the client, employer, community, social facilitator, Community liaison officer (CLO) and the Occupational health and safety (OHS) consultant. Technical specialists comprised the professional team and contractor, including the sub-contractors, who worked collaboratively through both cycles of the investigation to bring about value based outcomes.

The processes undertaken in cycles 1 and 2 provided an indication that the approach adapted from an integrated approach for Value Management was employed as one of the project management strategies. The approach contributed to project risk management on construction projects with the ability to resolve some of the identified risks causing project delays and also for the purpose of completing such projects within time, cost and quality. This was done in respect of internal or societal risks as other risks which had an equal potential to affect project success were not considered in this study (Al-Yami, 2008). The processes employed in this study which were adapted from the Integrated Approach for Soft Value management are planning, learning, analysis, creativity, evaluation and development stages which were undertaken during each of the stimulating action introduced to project activities presented in the previous chapter and also in the preceding sections of the current chapter (Al-Yami, 2008). Some elements derived from the proposed framework for value management depicted in Figure 2.6 denote a process starting with the organisations' mission and strategy. In this case, the mission was to improve the success rate of the construction project, which was behind schedule with a cost over-run (Al-Yami, 2008).

This process within the framework for value management was followed by among others, interface relationships and performance measurement systems within this study denotes the workshops conducted with focus groups and the project success criteria which is based on time, cost and quality. The next process is the continuous improvement which is denoted by the introduction of stimuli to project activities in order to improve processes for the purpose of achieving customer success, the last of the processes within the proposed framework for value management presented in Figure 2.6 (Al-Yami, 2008).

4.4.2   Sub-Question 1
'What are the methods of procurement for school construction projects?'

During the earlier introduction and literature review, scholars reported that the public procurement system of capital infrastructure projects is managed through the National Department of Public Works, which essentially prescribes procurement processes mainly done through the bid system, by means of closed and open tenders (Kaseke, 2011). This entails a procurement system in which a tender is advertised in public and specifications for compliance are prescribed where a successful tender is awarded from the complying submitted bids (May and Govender, 1998). It further suggests that procurement by open bid is carried out for tender classified as 90/10 in relation to tenders of a value exceeding R1m. In terms of closed tenders of bids below the R1m thresh-hold, also categorised as 90/20, three bidders are selected from the central database system of government and these bidders are invited to submit quotations based on the prescribed criteria and the bidder who best suits the criterea is awarded the project (Walker and Hampson, 2003; Bolton, 2007).

In response to sub-question 1, relating to the methods of procurement for construction projects, the study concluded that the procurement method of school construction projects was by public tender and the bid documents submitted by the successful bidder were used as the contract documents to regulate the relationship between the client and the bidder. The procurement method for sub-contractors on school construction projects is done through the close tender system by both the client and the successful bidder by way of nominated subcontractors and local subcontractors respectively (Bolton, 2007). Procurement for the main contractor on the school construction project under investigation was also done by open tender and the tender documents formed the basis of the contract between the client, C1 and the contractor, C8.

4.4.3   Sub-question 2
'What are the threat risks faced by each stakeholder?'

In terms of the threat risks faced by each stakeholder, the study through the various processes undertaken such as literature review and both cycles of the investigation revealed a series of threats in form of risks with which each stakeholder was faced. First of all, the study identified the stakeholder categories as, the client, the professional team, the community and contractors. These categories were found to have diverse but common risks such as;

4.4.3.1 Client

The risks pertinent to the client category, which includes the employer were identified as civil unrest, protest action, construction time, natural disasters, abandoning of the project by the contractor, poor quality work, uncertain productivity, weather or seasonal implications, financial risk such as inflation, wrong estimates, bureaucracy and fluctuating exchange rates (Skitmore, 2003).

4.4.3.2 Professional team

Risks common to the professional team which includes engineers, architects and quantity surveyors are, lack of coordination, constant scope changes by the client, lack of details in the concept by the project brief, use of wrong tools and equipment, design problems and delays in project implementation (Walker & Hampson, 2003; Bolton, 2007).

4.4.3.3 Community

The risks common to the community were identified as environmental (air, water and soil pollution), noise, health, injuries and delayed payments by the contractor and sub-contractors (Chow, 2007).

4.4.3.4 Contractor

The category of contractor which includes the main contractor and sub-contractors (nominated and local) were safety hazards, managing change orders, incomplete drawings, design errors, poorly defined scope, unknown site conditions, poorly written contracts, unexpected increase in material costs, labour shortages, damage or theft to equipment and tools, inclement weather, sub-contractor problems, poor project management and non-availability of materials (Walker & Hampson, 2003; Bolton, 2007).

4.4.4 Sub-question 3
'How are these risks currently managed?'

The risks identified in sub-question 2 are currently being managed as follows;

4.4.4.1 Client related risks

The risks pertinent to the client which were identified are currently managed by the provision of surety by the contractor to the client and also the deduction of 5% retention on every certificate submitted for payment to the client by the contractor.

4.4.4.2 Professional team

Risks related to the professional team are currently not being managed in any way apart from issuing the contractor with revised drawings where there is an error on the initially issued drawings.

4.4.4.3 Community

Currently, there is no management plan for the risks associated with the community

4.4.4.4 Contractor

In terms of contractor related risks, a provision for float and contingencies has been made in the current contract to manage the identified risks. Quality assessments are conducted by respective members of the professional team who inspect the work after every activity.

4.4.5 Sub-question 4
'How effective is the current project risk management?'

The description and status of the project site provided in chapter 1 in which it was reported that the project was behind schedule and faced with several challenges suggests that the current project risk management is not effective.

4.4.6 Sub-question 5
'How can Value Management be applied to PRM to improve project success on construction projects?'

The processes undertaken during both cycles of the investigation provided a possible way in which value management as a project management strategy could be applied to improve project success on construction projects. For the purpose of the current study, the researcher, started by conducting a study by analysing the available knowledge on the subject generated through the review of literature in order to identify some of the common problems experienced on construction sites. This usually results in the inability to achieve project success based on the project success criterea of project completion within time, cost and quality. The literature review was followed by the first phase of the cycle 2 investigation in which workshops were conducted with selected role-players on the project to isolate problems which were specific to the construction site upon which this study was based. The identification of project specific problems which were also referred to as industry risks led the researcher to make a comparison the problems highlighted in the literature review and consolidated a list of common causes identified by both processes. Data generated from the workshop participants and conclusions drawn from the literature review were compared in which common features were listed to form the basis upon which stimulating action was conceptualised and introduced to project activities for the purpose of recording and analysing the outcomes for interpretation.

The results indicated that the project management strategies employed on the project eradicated most of the problems or risks which were identified as major causes of delays on construction projects. It was also observed that after the implementation of project management strategies, the project schedule was restored to the newly prescribed timelines and no further delays were experienced thereafter. The results indicated that the processes undertaken in this study could be used as one of the project management strategies and applied to improve project success on construction projects.

4.4.7 Sub-question 6
'How effective would a Value Management contribution be to Project Risk Management?'

Sub-question 6 was addressed during the tenure of investigation cycles 1 and 2 in which the integrated approach for value management was adapted to suit the circumstances of the project under investigation. Cycle 1 investigation considered the various risk management tools and techniques currently in use such as FAST, FST, Decision trees, Contingencies and Float. An analysis of the literature review concluded that the contribution of these techniques to project risk management has not been effective because evidence available indicate that the techniques did not achieve the desired results in projects where such techniques were employed. Extensive studies further concluded that despite the application of the these techniques, the delivery of successfully completed construction projects continued to lag behind on projects where these techniques had been employed (Mahendra et al., 2013). This conclusion is also suggests that the constraints of time, cost and quality rank high among the causes of project failure because an effect on one of the constraints affects the others (Al-Yami, 2008). Conclusions drawn from literature caused the researcher to purposely exclude the use of contingencies and float from the proposed techniques which formed the stimulating action because contingencies and float were already being used as risk management strategies on many projects. Sub-question 1 was further dealt with in terms of the view that risk management was a structured approach to enumerating and understanding risk and in accordance with the definition of another scholar who described it as a formal orderly process for system identification to obtain acceptable control during the project lifecycle (Al-Bahar and Crandall, 1990; Baloi and Price, 2003).

The cycle 2 phase of the investigation involved the introduction of stimulating actions to project activities using procedures adapted from the value management processes, value management frameworks and the integrated approach for value management. All of these are found within the project management strategies domain which indicates that the use of project management strategies was more effective in managing risk comparison to the sole use of contingencies and float. It was further argued that in isolated cases, contingencies and float only catered for time and cost related problems and ignored quality. Quality is an equally important constraint on a project because projects of a poor quality, whose standards do not meet the minimum specifications are deemed not to be successful despite such projects being completed on time and within cost (Al-Bahar and Crandall, 1990; Baloi and Price, 2003). It is only in rare circumstances that projects completed beyond schedule and with a budget overrun are considered as successful. The point in question relates to the Sydney opera house and the River Thames barrier projects which were completed at 400% cost over-run and at a schedule three times more respectively to complete (Al-Bahar and Crandall, 1990; Baloi and Price, 2003).

In view of the interventions on the project activities derived from literature review which states that action research entails the analysis of the direct interventions of the researcher with the aim of analysing the outcomes, the researcher used stimulating actions as interventions to analyse project outcomes (Al-Yami, 2008). It can also be stated that this value-based design implies that participants' views are respected, they are treated as equals, and they participate freely and fully in the investigation. The participants are involved in looking into the situation that needs to be changed, thinking of solutions, and acting together to implement solutions to bring about change because the compliance of participants to the set standards by the researcher was a requisite in order for the desired outcomes resulting from the interventions to be achieved (Al-Yami, 2008).

While there are numerous risks which affect project success, some of which are internal while others are external, this study focused on internal risks and used a few intervening actions to some project activities in order to analyse the outcomes of such interventions. The results indicated that the stimulating action introduced to the selected project activities achieved the desired outcomes and in all the instances, the problems experienced on the project prior to the implementation of these project management strategies were resolved and no longer had the potential to negatively affect project objectives.

It was concluded that the project management strategy of integrated approach for value management proved to be an effective project management strategy for risk management on construction projects than the commonly applied techniques of contingencies and float. It was also noted that inspite of the foregoing, contingencies and float remain most common risk mitigation techniques used in construction projects.

4.5   DISCUSSION OF THE FINDINGS

The focus of this study was to review the current risk management tools for construction projects and the implementation of project management strategies. In relation to the problem being investigated of the backlog in the successful delivery of school infrastructure projects, numerous versions were presented through literature as the probable causes of delay and the various management tools, with their respective shortcomings were also highlighted. The study findings confirmed that many construction projects especially those managed by black contractors who are also categorised as Previously Disadvantaged Individuals (PDI's) were the most affected in the problems relating to the slow pace of delivery of construction projects because most of these construction companies were faced with problems ranging from lack of cashflow, lack of financial management skills, lack of experience and business management abilities.

The study also found that the traditional risk management techniques and tools currently used on construction projects such as, float, contingencies, brainstorming, Case Based Approach (CBA), Checklists, FAST Diagrams and FST were not bearing desired results in the curbing of problems which hindered the successful completion of construction projects. The study results further indicated that the use of an Integrated Approach for value Management was a worthwhile project management strategy to use on construction projects in order to maximise the value added outcomes and improve project success within the success criterea of time, cost and quality. The study also found that the interventions could not only be restricted to the ones used in this study but could also be extended to other areas of importance in accordance with dynamics of individual projects. The results further revealed that the implementation of the techniques within the project management strategy was not restricted to any specific phase of the project but could be implemented during any project phase based on the problem source or the area from where risk emanates excluding the phase of project closure.

The findings of the study further isolated one important aspect in the implementation of this project management process which was the involvement of all project stakeholders in an interactive process in which this study used focus group workshops to enhance the interactive part between the researcher and stakeholders. This process was very critical to the study because the researcher not only relied on the conclusions drawn from the literature review but also relied heavily on the experiences of the different stakeholders and role-player. The researcher acknowledged all views presented by participants regarding the perceived causes of delay which has the potential to hinder the successful completion of the project. The study also concluded that there was no specific sequence that needed to be complied with in terms of the implementation of the stimulating action. It was therefore suggested that there was need to study the dynamics of the project and obtain input of other stakeholders in order to objectively verify the causal effects of the highlighted project risks and to develop common interventions which would be supported by all role-players and stakeholders.

4.6 SUMMARY OF FINDINGS

This chapter presented results and attempted to assign meaning to the findings collated from the results in the preceding sections. The discussion section also analysed the findings and further tackled the processes used to address the research question and its corresponding sub-questions.

In order to appropriately discuss the results, it is imperative to reconsider the research aims and ascertain how the research questions related to the overall objectives of the study which prompted the researcher to explore the main research theme further. The central research theme which was explored in this study was phrased as;

"Reviewing risk management tools for construction projects and the implementation of project management strategies".

For the purpose of giving effect to the study, the above theme was translated into the main research problem which was presented as;

"Why do construction projects experience problems which impede their successful completion and what project management strategy can be used to effectively manage risk in order to improve the success rate".

The above problem statement further led to the development of the main research question which was set as *"Can an Integrated Approach for Value management be used as a project management strategy for risk management on construction projects?"*, the question was later broken down into sub-questions to effectively answer the most pertinent part of this study reflected in the problem statement which are; (i) *"Is the Integrated Approach for Value*

Management a more effective risk management technique for construction projects in comparison with the use of float and contingencies?, and (ii) "How can the Integrated Approach for Value Management be applied to enhance the ease of managing risk on construction projects".

The results which were derived from the investigation undertaken in two cycles explored the risk management tools currently in use and also conducted action research which revealed that float and contingencies were two of the many risk management tools currently used in construction projects and further concluded that other project management strategies such as the Integrated approach for value management was a technique worth considering when dealing with risk related problems on construction projects. Finally, the results revealed that the project management strategy employed in this study was a more effective risk management tool for construction projects in comparison with the traditional float and contingencies because more value was derived from the use of the former than the value derived from the use of the latter. The next chapter will present the conclusions and recommendations in which a detailed account of the conclusion will be discussed and also the recommendations arising from this study.

5.1 INTRODUCTION

In the previous chapter, the researcher dealt with the results and discussion thereof made on the basis of the topic investigated in this study. The research question which magnified the title into finer details was also highlighted. The study was conducted under the title, 'Reviewing of risk management tools for construction project and the implementation of project management strategies' in respect of risk management and value management. The study started by introducing the subject which was mainly centred on the backlog of school infrastructure which was reported to be caused by delays in the delivery of successfully completed projects. The study also explored the different risk management tools currently employed for risk management on projects and concluded that float, contingencies, decision trees, FAST models, FST, brainstorming, Case based approach and checklists were some of the tools and techniques in use despite the limited success rate (Mahendra *et al.*, 2013). The study also reviewed literature by different researchers who provided diverse views on the causes of delays in construction projects.

This study employed the epistemological approach within the critical theory in the ontological paradigm which propagated the use of a case study methodology using action research. The research process involved a two dimensional approach in which the investigations were conducted in terms of cycles 1 and 2 respectively. Cycle 1 involved the analysis of information sourced from literature while the cycle 2 investigation involved the interface between the researcher and project role players through focus group workshops in which participants were drawn from project stakeholders. The workshops provided data relating to the perceptions and views of the participants on the causes of delay on the project. Data generated from both processes were gathered, analysed, interpreted and conclusions drawn.

5.2 SUMMARY OF THE FINDINGS

The main aim of this study was to review the current risk management tools and test the application of the Integrated Approach for Value Management to manage project risk with the objective of increasing the project success rate based on the identified project success criteria and deliver projects within time, cost and quality to prevent further backlogs on construction projects. The project site selected for this study was a school construction project within the Ngaka Modiri Molema District Municipality of the North West Province at which the researcher was Project Manager. The following is a summary of the six chapters that made up the research report, and they outline how the participants and I addressed the research questions.

5.2.1 Chapter 1

Chapter 1 presented the rationale of the study in which the researcher argues for the need to answer the primary research question of *"Can an integrated approach for value management be used as a project management strategy for risk management on construction projects?"*. The main aim of this study was to review the current risk management tools and isolate the effect of float and contingencies on construction projects. In addition, the study was meant to compare the two with other project management strategies in terms of their effect in resolving problems which were identified as risks responsible for project delays which led to school infrastructure backlogs. The researcher interactively worked with project role players and stakeholders on the project to:

- Get their perception of the causes of delays on project activities.
- Explore the participants' proposals on how to improve the project success rate.
- Propose theoretical guidelines on how to ensure that the respective activities of each role player did not lag behind.

In order to provide further guidance to the study, the researcher utilised the following sub-questions; *"What are the methods of procurement for school construction projects?"*, *What are the threat risks faced by each stakeholder?*, *How are these risks currently managed?*, *How effective is the current project risk management?*, *How can Value Management be applied to PRM to improve project success on construction projects?*, *and How effective would a Value Management contribution be to Project Risk Management"*. The researcher also positioned these questions within the paradigmatic choices and justified the choice of action research as the research design for the study.

5.2.2 Chapter 2

Chapter 2 presented a literature review of the knowledge-base which forms the theoretical foundation of the study. In this chapter, the researcher provided an outline and established that although value management was widely known by most project practitioners, it was rarely used on construction projects (Pemsel and Wiewiora, 2013). The chapter also presented the different definitions of relevant concepts used for this study such as value management, value engineering, risk management, action research and also presented the theoretical framework on which the study was based. Further, the chapter presented the synthesis of value management and the different processes and procedures used for value management, risk management and action research. The project success criterea upon which this study was based were also presented as time, cost and quality in which an emphasis was also placed on the fact that despite the various project risks which may have an effect on project success, the focus of this study was confined to internal risk sources which were also referred to as societal risk sources. The literature consulted also revealed that while researchers argued that project success criterea is measured by time, cost and quality, other writers contradicted this view and stated that some projects completed over budget and beyond schedule would still be considered as successful such as the Sydney Opera House and the River Thames Barrier (Al-Bahar and Crandall, 1990; Baloi and Price, 2003).

An emphasis was also placed on the importance of stakeholder management as propagated by the Project Management Body of Knowledge which emphasised that stakeholders were people or entities who affected the project or those affected by the project and such people needed to be identified and their expectations managed (PMI, 2015). This stems from the fact that processes need to be in place to ensure that the end product of the project is what the client or customer expects because in certain cases, there are projects which have previously been undertaken without much stakeholder engagements resulting in no value add to the user-client or the intended customer. Some of the projects where stakeholder engagements were not effectively conducted are the Cape Town Stadium, the North Street taxi rank in Johannesburg and the e-tolls project on the Gauteng Freeways (Al-Bahar and Crandall, 1990; Baloi and Price, 2003).

Most of the reasons cited by the focus groups in the workshops as the main causes of delay matched with the reasons extracted from the works of various scholars and this served as a validation for the common issues highlighted by both sources led to the conceptualisation of the stimuli introduced to project activities and observed for analysis. In line with the above, the workshops revealed that the high staff turnover on the part of consultants made it difficult for an easy continuous flow of task execution because different consultants came with different instructions for the same task and this caused a delay in the process for the contractor to change the mind-set and adapt to the new way of executing the task hence the need to sign an MoU to maintain consistency.

5.2.3 Chapter 3

In Chapter 3, the researcher explained and justified the methodology and methods employed by the researcher for this study. The researcher used critical theory as the epistemological paradigm and participatory action research as the methodological paradigm in which he purposely chose to work with role players and project stakeholders because the entire project was behind schedule on account to tasks and activities which were executed by different role players which were also behind in schedule. The introduction of stimulating action to project activities was proposed in the design which was to be employed within action research.

The decision to use the chosen design was based on the fact that project management strategies needed to be implemented on the project site in order to change disposition of the project from being behind schedule and under budget to a state where it was on time and within budget following the approval of extension of time with cost by the client through employer. The researcher also preferred this methodology to allow participants to become fully involved in the improvement of the execution of their own activities within their roles on the project. This chapter also provided an outline of the methods used to ensure validity and ethical concerns in relation to the study.

5.2.4 Chapter 4

Chapter 4 addressed all the questions presented as; *Why do construction projects in experience problems which impede their successful completion and what project management strategy can be used to effectively manage risk in order to improve the success rate, Can an Integrated Approach for Value management be used as a project management strategy for risk management on construction projects?, Is the Integrated Approach for Value Management a more effective risk management technique for construction projects in* comparison with the use of float and contingencies and How can the Integrated Approach for Value Management be applied to enhance the ease of managing risk on construction projects respectively. The study also addressed and also explained what been had achieved in terms of the aim and objective which were listed as, to review the current risk management tools and test the application of the Integrated Approach for Value Management to manage project risk and to increase the project success rate based on the identified project success criteria and deliver projects within time, cost and quality to prevent further backlogs on construction projects respectively.

Based on the findings of action-research conducted by the researcher, five themes emerged, namely (1) the high staff turnover amongst the professional team (2) late delivery or wrong order of materials and also the procurement of sub-contracted services to the project from unskilled and incompetent sub-contractors (3) lack of skills and experience among the purported skilled labour of the contractor and sub-contractors to effectively support the listed compliance requirements for the project work specifications (4) lack of a proper stakeholder management process (5) lack of continuity which resulted necessitated the need for team induction and (6) Lack of the ability to enforce the contract penalty clause where some role players were in default. The second phase of the investigation which was referred to as cycle 2 dealt with the summary of stimulating action introduced to project activities which were summarised as; (i) The signing of a Memorandum of Understanding (MoU) with the professional team, (ii) Implementation of a procurement plan for material, plant and sub-contractors (iii) An audit of skills, trades and competencies, The Implementation of a stakeholder management plan, (iv) Team induction and (v) Contract Management, which entails the enforcement of the penalty clause.

5.2.5 Chapter 5

Chapter 5, which is the current and final chapter presents the conclusion and recommendations. The conclusion reviewed the entire process undertaken for this study from chapter 1 to the current chapter. The second part of this chapter reflects a discussion of the recommendations of this research and other concluding issues which are presented below.

5.2.6 Limitations of the study

The single case design is a limitation as it has a tendency to constrain the outcomes which have a tendency to be subjective and based on the dynamics unique to this project. Therefore, the results of this study are only limited to the construction project site being investigated and may not be generalised and applied to other construction sites.

5.3   CONCLUSION

The findings from the two research approaches were used to draw conclusions which were also enhanced by an additional post action research interaction with participants where the outcomes were evaluated and conclusions made. The post action research interaction with role players revealed that stakeholders perceived the implemented project management strategy as a success. In addition, the participants indicated that if the strategy had been employed at the start of the project, all the project activities could have been completed within the initial timeframe. All participants indicated that the interventions made altered the project perception, as most of the problems and challenges previously experienced had been resolved and the site operations were without any hurdles. The respondents further indicated that if the above processes had been implemented during the early phases of the project, all the risks that had previously contributed to the delays would be identified and managed, which could have restored the project back on schedule. Further, the participants also recommended the need to implement the strategy used in this study on every project.

5.4   RESEARCH LIMITATIONS

The limitations of the study were summarised by four factors as reflected in the following discussion:

5.4.1 Project size

The project upon which the study was based was a relatively medium project where the findings were restricted to the referred to project and not to be generalised to other projects.

5.4.2 Limitation in the number of respondents

In view of the fact that the project forming the basis of this study was classified as a relatively medium project, only 15 participants drawn from project stakeholders were chosen in the embedded survey within the case study design. This was considered as a limitation as the findings may not be generalised to make an inference to other projects. The findings of this study are limited to the sample of 15 participants from the project site which was investigated where the project was fraught with many challenges which resulted in the delay of most activities.

5.4.3 *Geographical location of the project*

The geographical location of the project was a limitation which restricted the findings from being generalised to other projects because the project upon which this study was based was located in a rural setting with very high poverty and unemployment levels. The high poverty and unemployment levels within the surrounding communities had an effect on the project because economic and social conditions compelled members of these communities to protest and demand for economic benefits from the project. This was cited as a limitation because the social and economic dynamics of communities in areas where projects are located in economically empowered areas did not experience similar problems as those faced on the current project. Previous research concluded that people in less impoverished areas such as suburbs were not economically dependent on projects for survival and had less interest in participating in protests against the projects (Akintoye and MacLeod, 1997; Booysen, 2007).

5.4.4 *Uniqueness of the project*

The unique characteristics of the project setting and the dynamics of the project were a limitation which could only be ascribed to the project on which this study was based because the team combination, client, customer, community and the type of project made it a unique project whose characteristics could not be replicated and generalised to other projects.

5.5 IMPLICATIONS

The study substantiated the need for a more versatile project management strategy with a diverse and multi-dimensional problem solving ability as a risk management tool and also took into account the methodological processes which required to be standardised as a project management model on construction projects. The study further identified the importance of establishing a system through which a comprehensive project management strategy would be included in all project documents during the planning phase as an addition to the usual float and contingencies.

5.6. RECOMMENDATIONS

This researcher recommended the following;

5.6.1 Incorporation of a VM clause into the project charter

It was recommended that a value management clause pertinent to risk be incorporated into every project charter. This was for the purpose of compliance with value management processes during the initiation phase of the project. The incorporation would contribute to early detection and prevention of risks on construction projects which would ultimately lead to an improved project success rate.

5.6.2 Documenting of findings

This study also recommended that findings and conclusions of similar studies be documented and kept in an information repository. This would ensure that the lessons learnt from the current project were available to future project functionaries who may experience similar problems on their respective projects. This recommendation was drawn from the review of literature which revealed that one of the reasons for project failure was a lack of reference to databanks for lessons learnt by project stakeholder (Baloyi and Bekker, 2011; Bowen, Cattell, *et al.*, 2010; PMI, 2015; Tah and carr, 2001). The researcher further states that once the findings are documented and archived, a comprehensive and standard model could be developed and packaged as a value management tool kit within the project management strategies for risk management on construction projects.

5.6.3 Addition of other management strategies to PRM

The study recommended that project management strategies should be added to construction contract documents to complement float and contingencies already factored into project documents. This is meant mitigate a variety of risks because project management strategies were better able to identify and deal with risk at any phase of the project unlike float and contingencies which were only applicable to mitigate cost and schedule related project risks.

5.6.4 Recommendations for future research

The study recommended further research to be undertaken in different aspects of the subject matter in order to develop empirical evidence upon which to base VM as a risk management tool. The fact that the field of value management is vast and wide shows that there is still potential for further research to enhance the knowledge currently available. This study further answered the question *"Why do construction projects experience problems which impede their successful completion and what project management strategy can be used to effectively manage risks in order to improve the success rate?"* However, there are several other questions which arose during the process and they need to be investigated further. For example:

5.6.4.1 Undertaking of a more representative study

A more representative study incorporating a substantially large number of respondents within professionals in the construction sector and the use of many diverse construction projects is recommended.

5.6.4.2 Repeat and Validation of the VM process

This study recommended that the VM process applied needed to be repeated and tested on more construction projects of different sizes and also increase the number of participants to more people within the construction environment to establish if the same outcome would be achieved.

5.7   CRITICAL REFLECTON

During the course of the researchers' interaction with role players, it also became apparent that some of the stakeholders on this project site were not comfortable with investigating their practice. They felt threatened by this exposure of their lack of knowledge and competence such that it took them a bit of time to gain confidence and fully participate in the process although the interaction was enough to generate data to answer the research questions.

In addition to generating these themes, the cycle of action and enquiry (Cycle 1) was an important learning experience for the researcher in his individual capacity. The critical reflection on the researcher in this cycle helped to realise that his own role as a researcher was not consistent with the values of action research. For instance, it was realised that instead of facilitating the workshops with the participants, the researcher directed them, dominating and contributing more than the participants during the discussions. However, the learning acquired by the researcher facilitated corrective action which was implemented by the researcher and the revised approach during the first cycle informed the steps and approaches that were used in the second cycle. The researcher further learnt to work more collaboratively with the participants. Chapter 4 further emphasised the importance of critical reflection on the part of the researcher and the need to turn negative experiences into opportunities to learn, which is consistent with the principles of action research.

5.8   SUMMARY

The aim of this study was to work collaboratively with project stakeholders on the school construction project for the purpose of reviewing and implementing a project management strategy to improve the project success rate. The purpose was to fill the gap that was evident from the literature. Literature suggested that the current risk management tools and techniques were hardly capable of mitigating challenges faced by construction projects which have led to most projects being completed beyond schedule, cost and below the desired quality. The participants in this study were in a position to lead instructional tasks of the project, however, they had administration tasks that made it difficult for them to complete tasks or activities within the desired time, cost and quality. This study argued that stakeholders should take an active role as task executing leaders.

Guided by the primary research question, the researcher used democratic methodological paradigms and data gathering strategies to conduct this study. This was meant to allow participants to be actively involved in working out the improvement of their social context, without any undue external influence. The findings from Cycle 1 demonstrate that the respective participants were dependent on one another as some activities which were preceded by those which had to be executed by another participant could not be started before the participant responsible for the preceding activity completed that activity.

The findings from Cycle 2 demonstrated an improvement in the management of activities by eradicating potential threats to the successful completion of activities which led to continuous project success. The findings from this cycle also revealed the importance of collaboration, action research and reflection. Action research allows people to do things by themselves, at their own pace in a favourable atmosphere – a peaceful and relaxed milieu. Lastly, the findings of this study demonstrated the need for not only the stakeholders to be involved in the improvement of processes to attain project success but for all role players who are involved in the construction project to be involved in the whole project value cycle in order to achieve the desired outcomes by managing all risks which have the potential to negatively affect project objectives.